QUALITY ASSURANCE IN
RESEARCH AND DEVELOPMENT

INDUSTRIAL ENGINEERING

A Series of Reference Books and Textbooks

Editor

WILBUR MEIER, JR.
Dean, College of Engineering
The Pennsylvania State University
University Park, Pennsylvania

Additional Volumes in Preparation

QUALITY ASSURANCE IN RESEARCH AND DEVELOPMENT

George W. Roberts

Babcock & Wilcox
A McDermott Company
Alliance, Ohio

CRC Press

Taylor & Francis Group

Boca Raton London New York

CRC Press is an imprint of the
Taylor & Francis Group, an **informa** business

First published 1983 by Marcel Dekker, Inc.

Published 2019 by CRC Press
Taylor & Francis Group
6000 Broken Sound Parkway NW, Suite 300
Boca Raton, FL 33487-2742

First issued in paperback 2019

No claim to original U.S. Government works

ISBN 13: 978-0-367-45189-9 (pbk)
ISBN 13: 978-0-8247-7071-6 (hbk)

Visit the Taylor & Francis Web site at
http://www.taylorandfrancis.com

and the CRC Press Web site at
http://www.crcpress.com

Library of Congress Cataloging in Publication Data
Roberts, George W., [date]
     Quality assurance in research and development.

     (Industrial engineering ; v. 8)
     Includes bibliographical references and index.
     1. Research, Industrial--Management.   2. Quality assurance.
I. Title.   II. Series.
T175.5.R6   1983   658.5'7'0287   83-15328
ISBN 0-8247-7071-4

To my wife, Sara Ray

This book is intended to provide guidelines for obtaining research and development results of a consistent and known quality. Some traditional quality systems originally developed for hardware production are presented and their modification for use in a research environment discussed. Systems which have been designed for management of the research and development function are also presented as tools for the assurance of R&D quality.

The approaches in this book were developed for an industrial research center engaged in R&D to support operating divisions of the parent corporation. Hence, the discussions on annual R&D plans and the use of technical coordinating committees. This ensures constant communication of the needs of the operating divisions in response to market demands for new products or advanced technology.

Contracted research puts an added strain on standardized administrative systems because each customer has differing needs and expectations. For this reason much of the text is directed to tailoring the standard QA system by using levels or individual project QA plans.

For all the emphasis on applied R&D, the systems discussed herein are just as applicable to basic research. This is true either from the perspective of reducing cost and increasing reliability of complex test apparatus or for assuring that phenomena are observed in a properly controlled and reliable laboratory environment. The material was developed primarily from the author's experience and information available through other research centers and available literature. It represents the author's opinion of management practices that are appropriate for implementation at any research center.

The author assumes the reader has some familiarity with basic principles of quality assurance or "total quality control." It is hoped that the text is equally useful to a research specialist or manager setting up a QA system within a research center or to a QA professional making the transition from production line to the research center. This text is not intended to supplant any of the major works on quality assurance or quality control. It should be used as an adjunct or an interpretation of those texts for R&D.

The central theme of this book was first presented by the author in a paper for the American Society of Mechanical Engineers (ASME) and published in the September 1978 issue of <u>Mechanical Engineering</u> magazine. Substantial sections of that article are reprinted in this book, particularly in Chapter 7.

The author wishes to express his appreciation to the management of the B&W Research and Development Division. Not only the writing of this book, but also the success of the QA program has hinged upon their support. There are several support groups within the Alliance Research Center who have worked hard preparing art and editing manuscript. To them it may have been all in a day's work, but to me it was invaluable. Technical sections helped by providing examples from typical projects. But much more than that, they took the concepts that we jointly hammered out over the years and made them work on the job. The QA staff has all been very helpful by providing support, illustrations, comments, and understanding while I was locked up in my office "working on the book." This is their story most of all. A special thanks to Nina Thomas of the Lynchburg Research Center for her major contribution to Chapter 10 on Software Quality Assurance. Also to Howard Case for reviewing Chapter 10. I wish to thank the editors of the <u>American Society for Testing and Materials</u> (ASTM) <u>Magazine</u>, <u>Standardization News</u>, and Dr. Brian Belanger for their permission to use major excerpts from their article, "Traceability — An Evolving Concept," in Chapter 8.

Finally, I would like to acknowledge the support and patience of Madelyn Dunn who typed the manuscript.

George W. Roberts

QUALITY ASSURANCE IN RESEARCH AND DEVELOPMENT

Chapter 1

RATIONALE FOR A QUALITY ASSURANCE PROGRAM AT A RESEARCH CENTER

INTRODUCTION

Examining the rationale for quality assurance in research and development requires one to look at the reasons for R&D in the first place and to understand the importance of R&D in the life of the company. The success of a company is directly related to its productivity. Increased productivity permits companies to give more value per dollar to its customers, to operate more efficiently internally and to thereby increase sales and profits. The ability to increase productivity is directly related to fields having foundation in science and technology — the direct result of organized research and development. Therefore, R&D is the key link in the productivity chain (1).

The advantages for a sound QA program in the R&D activity stem from four areas of concern: product liability, government regulation, sales, and costs. Development testing provides preproduction data that characterizes the product's design. It establishes product performance parameters that form the basis for design tradeoffs for reliability assessments, cost optimization studies, and product liability risk assessments. The results of R&D are often part of the justification for a new or changed design. Manufacturers are being held to the level of experts in their field by the courts (2). If R&D provides the confidence that the design is safe, then the R&D will be subject to scrutiny by the courts in deciding to what extent a manufacturer was reasonably prudent in providing protection to the public. Therefore, the records to substantiate the conclusions and recommendations from an R&D project should be both available and understandable.

Certain government specifications and federal codes require extensive testing to prove the capability of a product to function under adverse conditions. This qualification testing, whether done for reliability assessments or for safety assessments, is required to be done under carefully monitored conditions.

Properly qualified R&D provides substantiated data to back up performance claims to customers and can be used as an effective tool in

1

promoting sales. For high cost items, it provides added confidence to a customer that the item will perform as intended, particularly for first-of-a-kind or custom-built facilities or systems. The data then backs up the manufacturer's specification for maintenance and optimum operating conditions for the item. It provides a clear set of documentation to evaluate the reasons for success or failure of the item and can be used as a warning to customers to avoid certain undesirable practices.

Ultimately, the question comes down to a matter of cost. Constructing R&D test facilities costs money for the company. The company needs the confidence that those facilities will provide accurate data and meet performance standards established by the customer or its own research scientists and engineers. Also, by properly qualifying R&D with adequate systems and controls, the overall effort to qualify a product design can be reduced (3). Records from the R&D are capable of justifying specific design concepts without the need for complete qualification testing of production first articles. Certain design assumptions proven during R&D can be supported by the QA activity. Qualification testing or full-size prototype testing can then concentrate on proving production processes or system level design concepts which can only be verified by a completed first article test program.

Malcolm Baldridge, U.S. Secretary of Commerce, stated that,

> For managers the challenge is to create an organizational
> environment that fosters creativity, productivity, and
> quality consciousness (4).

He pointed out that 40% of all costs in getting a product to the marketplace are in the design cycle. He also stated that top management must emphasize prevention, rather than correction. This is a recurring theme with all experts in the field of quality management. The National Advisory Council for Quality identified eight universal quality improvement steps (5). Number 7 states that emphasis for quality must be shifted from error detection to error prevention. This is the same principle used for navigational correction. The earlier you make the change, the greater the effect down the road. But, obviously, you need good information before you make decisions to change. Money spent assuring the quality of research data will pay dividends over again when the product enters the marketplace.

Crosby said,

> Quality improvement through defect prevention...is the
> foundation of all ITT quality programs (6).

In their discussions on the cost of quality, Juran and Gryna (7) identify a company's activities and costs as:

1. Costs of market research to determine customer needs.
2. Research and development costs to create improved product concepts.
3. Design costs to translate concepts into product specifications.

4. Costs of manufacturing planning.
5. Costs to maintain precision of machines and processes.
6. Costs to operate process controls.
7. Costs to market the product's quality.
8. Costs to appraise the product and measure its conformance.
9. Costs of defect prevention.
10. Losses due to unquality.
11. Cost to inform plant management and personnel about the status of the quality function.

Of these, the first three are concerned with determining, creating, and defining fitness for use. The remaining deal with conformance to specification. In this context it would be a mistake to define the total quality improvement effort as the achieving of conformance to specification since that would ignore the crucial first three activities that are making sure the specification was right in the first place. When designs are rushed to production without proper verification or based on technology derived from hastily conceived and executed experiments, the stage is set for a long and costly process of piecemeal defect detection, analysis and feedback for design correction actions that could have been prevented. If the company has a research and development organization, that company should heed the National Advisory Council for Quality's advice in universal quality improvement step Number 4, "Formal quality improvement activity must be launched in every organization."

George Hardigg, Vice-President and General Manager of Westinghouse's Advanced Power Systems Division, put it very succinctly when he said,

> ...Studies and experiments that are not conducted under controlled verifiable conditions and thereby produce data that is useless in supporting subsequent design decisions are a waste of time and money (8).

REFERENCES

1. Hughes Aircraft Company, R&D Productivity, Hughes Aircraft Company, Culver City, CA (1978).
2. Product Liability, The Present Attack, American Management Association, p. 24 (1970).
3. G. Roberts, Quality Assurance in R&D, Mechanical Engineering, Vol. 100, Issue No. 9, p. 41 (1978).
4. M. Baldridge, Designing for Productivity, Design News, Vol. 38, No. 13, p. 11 (July 1982).
5. News & Trends, Production Engineering, Vol. 29, No. 7, p. 8 (July 1982).

6. P. Crosby, Quality is Free, McGraw-Hill, New York (1979).
7. J. M. Juran and F. M. Gryna, Jr., Quality Planning and Analysis, McGraw-Hill, New York (1970).
8. G. Hardigg, Proc. Sixth Annual Nat'l. Energy Div. Conf. ASQC, EP1.17 (1979).

Chapter 2

PLANNING THE OVERALL
QUALITY ASSURANCE PROGRAM

I. GETTING STARTED

A company that is successful is satisfying its customer's requirements and does in fact have some sort of quality assurance program, although it may not be nearly as effective as it could be. Most companies do define their quality standards, although they might not recognize them as such. The standards might be called policy manuals, standard operating procedures, or administrative procedures, but if these documents provide a standard mode of operation or conduct, they in fact constitute a quality standard.

To begin to establish an effective quality assurance program, a research center should collect and review data about its existing activities. This process in itself may point out deficiencies in operation or an unclear or conflicting direction that will require correction. Goetz (1) suggests the following steps:

1. Collecting data concerning existing systems and procedures.
2. Preparing process flow charts depicting in detail the activities that are actually being performed.
3. Preparing written descriptions of these basic systems.
4. Correlating your company's needs and government regulations with your actual practices.
5. Modifying systems and procedures to satisfy all of these needs.
6. Monitoring the new or revised systems to ensure their continued effectiveness.

Top management must examine the type of pressures it is receiving from outside the center. Is the QA program desired to reduce costs or risks of liability, or must it address specific requirements levied by customers or regulatory agencies?

Federal requirements which affect the research process are delineated in the NASA NHB 5300 series specifications (2), Department of Defense

MIL-Q-9858A (3), Appendix B to 10 CFR 50 (4), and <u>ANSI/ASME</u> NQA-1
(5), to name a few. If more than one of these are likely to be levied upon
the research center, those components which are common to all of the specifications (e.g., Control of Measuring and Test Equipment) should form the
baseline for the research center standard quality systems.

As can be seen in Table 2.1, many program elements are quite similar from one specification to another. They, in turn, are very similar to
the basic elements of quality disciplines described in commercial quality
control applications described by Feigenbaum (6), Juran (7), and other
authors. As an aid in determining what the common components of the various specifications may be, particularly in a nuclear environment, the
reader may wish to review another matrix of nuclear QA requirements published by the American Society for Quality Control (ASQC) (8). In this matrix, 71 quality assurance elements are extracted from the criteria of 10
CFR 50, Appendix B, and are compared with the requirements from five
other quality program documents. Similarly, the same 71 quality assurance elements are compared with an additional 10 consensus standards.
Those areas contained in specifications which are not likely to be levied
upon the research center can be left until later, provided the research center has included within its system a means for adequately handling special
requirements during the detailed planning phase for each project. Additional guidance on how to address special requirements not previously addressed by the research center's standard systems will be described in
Chapter 4.

Management should also determine the need for compliance with state
building codes and environmental requirements for the research center. If,
as a result of local building codes, there is a necessity for qualification of
welders for installing facility piping or pressure vessels, the requirements
for control of such facility work can, if desired, be incorporated into the
overall research center QA program.

Finally, the research center should identify its internal needs for a
QA program for the research process. Among those needs might be the
desire for improved accuracy of research results, cost reduction, pressures of competition of other research laboratories, or certification to
meet national accreditation standards.

Some managers may accept some features of a classical QA program
but not others due to the background and experience of the manager. If the
manager has seen the advantages of an inspection system, then QA might
mean inspection. Implementation of that phase of the QA program will receive solid support up the management line. Other QA elements may not
be accepted as easily. The program may have to infringe on some special
prerogatives of certain organizations. Management must understand and
agree what the QA program is and what it is going to do for the center —
before they launch the program.

Table 2.1 Matrix Quality Assurance Requirements

TITLE	10CFR50 APP B	MIL-Q 9858A	ASME NA4000	ANSI N45.2	RDT F2-2T
ORGANIZATION	I	3.1	4210	3	2.3
QUALITY ASSURANCE PLAN REVIEW AND APPROVAL OF PLAN	II PSAR/FSAR	3.2 1.2	4111 4112, 4120	2 NOT SPECIFIED	2.2 2.2.2
DESIGN CONTROL	III	4.1	4410	4	3
PROCUREMENT DOCUMENT CONTROL	IV	5.1 5.2	4430 4441	5	4
INSTRUCTIONS, PROCEDURES AND DRAWINGS	V		4140	6	2.4
DOCUMENT CONTROL	VI	6.2	4420	7	5.3
CONTROL OF PURCHASED MATERIALS	VII	4.1	4430	8	3.4, 5.7
IDENTIFICATION AND CONTROL OF MATERIALS, PARTS AND COMPONENTS	VIII	5.1 6.1	4441 4442	9	4 5.4
CONTROL OF SPECIAL PROCESSES	IX	NO TRACEABILITY	4451	10	5.5
INSPECTION	X	6.2	4510 4520, 4530	11	5.6
TEST CONTROL	XI	6.3	4510 4520, 4530	12	5.6
CONTROL OF MEASURING AND TEST EQUIPMENT	XII	6.3	4600	13	5.8
HANDLING STORAGE AND SHIPPING	XIII	4.2, 4.3 4.4, 4.5 6.4	4460 4452	14	5.12
INSPECTION TEST AND OPERATING STATUS	XIV	6.7	4540	15	5.6.4
NONCONFORMING MATERIALS, PARTS OR COMPONENTS	XV	6.5	4550	16	5.10
CORRECTIVE ACTION	XVI	3.5	4800	17	2.6
QUALITY ASSURANCE RECORDS	XVII	3.4	4900	18	2.4
AUDITS	XVIII	NONE	4700	19	8

Source : From Ref. 16.

II. ORGANIZATION

Most organizations consist of a structure (and communication paths within
the structure) to direct the desired effort toward goals that the organization
intends to fulfill. The corporation's basic structure may contain a board of
directors, a president, corporate staffs, and operating arms. Line and
staff relations might be defined as:

1. Line authority invested in the president and operating arm or division
    managers.
2. Corporate staffs who represent company disciplines; i.e., manufactur-
    ing engineering, quality, accounting, etc. Although they possess no line
    authority, they do provide technical leadership, develop policies, and
    monitor implementation throughout the company.

   Within the staff, QA may be represented by a corporate member who
provides common direction to operating divisions in discharging quality
tasks. In addition, the corporate member requires the divisions to assure
the attainment of the required quality level for the product produced by that
division. Each division manager will then establish a functional definition
of quality through the appointment of a quality manager and issuance of a
QA manual specifically for that division. Obviously, it is the communications system that is necessary to properly serve a company functional discipline. In this relationship, communications would exist between the corporate staff and the president, the corporate staff and divisions, between the
corporate staffs, and between division functional managers and division
program managers.
   Other high technology companies have adopted this philosophy of des-
ignating "functional executives." TRW, for example, has appointed Vice
Presidents for Quality, Productivity, Manufacturing, Material and Telecom-
munications and Technical Resources (9).
   Within the division, the communication system must extend from the
top down. Particularly, in the research environment where much of the
quality control tasks are delegated to project leaders, the communication
of quality objectives and techniques must be thorough. The QA organization
must develop plans consistent with product and local requirements while being consistent with division and staff policies. It must develop and maintain
an effective organization with fully qualified personnel and strive to improve
the technical and administrative functions within the cognizance of the quality function. To accomplish this, the Quality Assurance organization must
have sufficient authority and responsibility for establishing, planning and
implementing the required quality program.
   Although no one organizational arrangement is preferable, some organized approach is necessary and should be demonstrated by appropriate
charts and written descriptions that clearly define the authority and duties
of all persons involved in the quality program. Quality personnel should

have sufficient organizational freedom to initiate, require, and verify resolutions to quality problems. This freedom should include sufficient authority to control further processing or testing until the nonconforming item, deficiency, or unsatisfactory condition has been properly resolved. This authority to control further processing should be extended to identify the conditions for a "stop-work" authority to be given to the quality organization (or to some other entity such as a review board).

Research projects often involve a few people full-time: a project leader, one or two engineers or scientists, and possibly one or two technicians. These individuals are backed up by various support services, such as machine shops and instruments groups. With working teams this small, it is inappropriate to include a full-time quality assurance professional to perform the various checking and verification activities. The technical expertise in research centers is extremely diverse and prohibits the paralleling of skills that is practiced by some production QA organizations in large companies. Therefore, it is necessary to delegate most of the assurance activities to the operating project organizations (10).

Quality Assurance (QA) can be viewed as a total management process. The overall assurance of the attainment of quality objectives is done through the use of not only the quality control organization, but through the influencing of activities of other organizations involved with or having an effect on quality. Quality Control (QC), on the other hand, is more directly involved in the actual physical control of the product, the hardware, or the activity. Persons responsible for quality control exercise direct in-line approval or measurement verification of the attributes of the items involved.

Responsibilities will vary from company to company, but in a research and development environment, the quality control functions are often delegated to support activities that are assisting project personnel. When the procedures for control of these support activities are prepared, the inspection, checking, and verification activities can be written into those procedures with the inspection and checking activities assigned to the organizations normally involved in those systems. For example, design checking can be delegated to the design organization as long as some independent and competent person is identified as the checker for designs or checker for drawings. This is a quality control function as opposed to the quality assurance function of verifying that the inspection activity is in place and being performed effectively.

The major role for the QA organization is overall systems development, project planning, training, and auditing to verify implementation of the inspection, checking and verification activities. This distinction between quality assurance and quality control is taken from the Appendix B to 10 CFR 50 (4). In that specification, quality assurance is defined as "all those planned and systematic actions necessary to provide adequate confidence that a structure, system, or component will perform satisfactorily in service." The term Quality Assurance is meant to include quality control, which is defined as "those quality assurance actions related to the physical

characteristics of a material, structure, component or system which provide a means to control the quality of a material, structure, component or
system to predetermined requirements. " In a sense, the point could be
made that quality assurance in a research center is more oriented to the
validity of data, whereas quality control is more oriented to test section
hardware compliance to drawing or specification.

    If the physical characteristics of the test section are critical to the
quality or validity of the data, then quality control functions must be imposed to assure conformance of that hardware to predetermined specifications. However, it is the overall goal of the quality assurance function to
cause the determination to be made of:

1. Whether the hardware is critical to the data or not; and
2. To verify that activities affecting the critical hardware <u>and</u> the data
    (e. g. , calculations, design of experiment, data reduction and reporting)
    are subject to verification activities.

    As stated before, these verification activities in a research environ-
ment would probably not be actually performed by the QA organization, but
would be delegated to the appropriate organization. Many quality control
functions such as inspection and testing can be delegated, but this is a decision that should be made on a cost benefit basis to the individual research
center. In either case, the adequacy of the quality assurance/quality control systems should be verified by the quality assurance organization
through a system of audits.

III. POLICIES AND PROCEDURES

A. Overall Policy

It is vitally important to the success of the quality program that it has the
support of top management. The first step in providing this support is to
generate a policy signed by the top executive stating that there will be a QA
program in force at the research center. It should be mandatory that the
organizations and personnel of that research center comply with the QA
program defined in the quality manual and implementing procedures. The
policy should identify the person in charge of the quality program — the person who will have the responsibility for establishing, defining, implementing, and enforcing the policies and the procedures to ensure compliance
with the program. The policy also should include the authority of the person
in charge and should spell out how the resolution of conflicts will be handled.

B. Subtier Policies — The QA Manual

Sublevel policy statements must identify the research center's actions to be
taken with respect to each area of activity. The QA manual is often a
series of policy statements containing more detail than the top level policy,

but not as detailed as the specific implementing procedures which follow the manual. Some manuals have sections identified as policy sections with each one given a unique policy number. There should be sufficient sections within the manual to address each element of work activity using the basic model specification as a guide. It is well to recognize that each research project will have some differences. The capability to be flexible in imposing appropriate quality requirements on each project must be built into this system from the very start. Therefore, it might be wise to write into the QA manual the capability for using different levels of quality (see Section IV).

The quality manual can be structured around any type of model specification. Using Appendix B to 10 CFR 50 as a guide, the list of elements to be included in the quality manual are:

1. Organization
2. Quality Assurance Program
3. Design Control
4. Procurement Document Control
5. Instructions, Procedures and Drawings
6. Document Control
7. Control of Purchased Material
8. Identification and Control of Materials, Parts and Components
9. Control of Special Processes
10. Inspection
11. Test Control
12. Control of Measuring and Test Equipment
13. Handling, Storage and Shipping
14. Inspection, Test and Operating Status
15. Nonconforming Materials, Parts or Components
16. Corrective Action
17. Quality Assurance Records
18. Audits

An alternate approach is the structure outlined by RDT F2-2 (11):

1. Introduction
2. Management and Planning
3. Design and Development
4. Procurement
5. Manufacturing, Fabrication and Assembly
6. Construction and Installation
7. Operation, Maintenance and Modification
8. Quality Assurance Audits

With these items establishing the topical format of the QA manual, it can then address each of these items as they apply to the research effort, describe how that research center operates, and include illustrations of the forms and systems employed by the research center.

C. Implementing Procedures

Procedures should be written by the people who are most affected by them —
shipping procedures by the shipping department, drafting standards by the
drafting group, and so forth. They then should be reviewed for compliance
with QA policy, routed for general review and then recycled as necessary to
resolve differences. One suggestion is to appoint a procedure coordinator to
obtain and follow-up on commitments for procedure preparation and review.
R&D people must recognize that it is possible to standardize methods used in
various disciplines — chemistry can use the same calculation sheets as metal-
lurgy. However, the QA manager should keep in mind that R&D activities are
extremely changeable and that the procedures and forms must be flexible
enough to accommodate these changing needs.

The QA manual and implementing procedures now form the research
center standard systems from which elements can be selectively drawn to
meet customer needs.

IV. QUALITY LEVELS — APPLYING THE SYSTEM

A research facility may be involved in many different research projects —
some of short duration and some over a period of years. Each of these projects require some degree of individualized planning to accommodate their
uniqueness. The planning process must be thorough but rapid to permit the
project to begin quickly. Many research projects are generated in support of
a problem which has arisen in the field or at a plant, and the investigation
must begin immediately — sometimes in a matter of a few hours. Despite
this, the potential impact of the research results on personnel safety or in
time and money for anticipated design or process changes dictates that quality
standards should not be put aside just to get the data out. On the contrary, it
is at this time when efforts to assure accuracy and reliability of data may be
most needed. The QA system for this type of environment must be capable
of responding to these needs. The project leader needs to spend as much
time on the technical details as possible and not be unduly hassled with an
unwieldy bureaucracy. With a properly designed system, the QA requirements will be appropriate to the technical needs of the data and applied
with a minimum of time and effort.

When deciding what quality systems need to be implemented, it is well
to remember that, in most cases, R&D effort does not involve manufacturing
or delivering of hardware. Hardware may be produced for an individual project, but the end result of most R&D effort is data. The quality program must
establish systems that are pertinent to assuring the quality and validity of the
data or results. Sections of a quality manual for a production QC program
may be quite specific as to how specific pieces are to be inspected, tested,
packaged, and shipped; but in an R&D QA manual, the question is always
whether these activities have a real effect on the validity of the research

data. Therefore, many of the quality elements may be imposed by a specific quality plan.

If the QA manual was able to be entirely preplanned to cover all instances, each section of the quality manual could be broken down into various quality levels with specific activities required by level. For example, in a section for "Inspection," Quality Level "1" might not require any inspection to be performed; Quality Level "2" may permit an inspection by the project leader or his designee; Quality Level "3" requires inspection by an independent quality control organization; and Quality Level "4" may require customer or government inspection activities.

An interesting approach towards varying quality levels is presented by the Canadian Standards Association in their Z299 Series of standards (12). Figure 2.1 shows various quality programs defined by the Z299 series. As the need for increasing levels of quality is established, additional elements are added by successive standards in the series. The progression from Category 4 up to Category 1 accumulates additional control elements until the complete set of requirements for nuclear programs is imposed at the Category 1 level. The Canadian Standards also provide some thought provoking

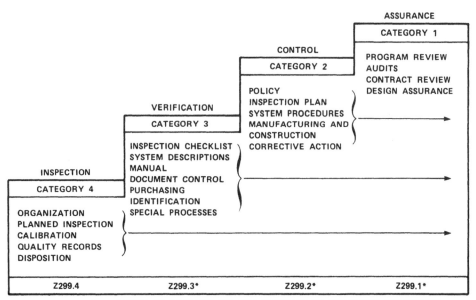

*EACH OF THESE Z299 STANDARDS CONTAINS ALL THE FEATURES OF THE LOWER CATEGORIES.

Fig. 2.1. Comparison of main features. (With the permission of CSA this material is reproduced from CSA Z299.0-1979, Guide for Selecting and Implementing the CSA Z299 Quality Program Standards, which is copyrighted by the Canadian Standards Association, and copies of which may be purchased from the Association, 178 Rexdale Boulevard, Rexdale, Ontario M9W 1R3.)

guidelines in the area of quality levels versus cost benefits. Many of us are familiar with Juran's Quality Cost Curve (7), as shown in Figure 2.2. The Canadians have depicted a more time phased illustration of quality costs showing the realities of starting a new program. Figure 2.3 shows there will be an initial loss due to inefficiency in defining and implementing the program. There is a subsequent break-even point, however, due to reduced costs associated with manufacturing defects, and service claims. These are classified as internal and external failures. The potential for a substantial initial loss should be recognized before embarking on any quality program. Figure 2.4 shows the traditional quality cost categories before and after implementing the quality program. Note in this figure that the internal and external failures are combined into one block. In an R&D program, the true costs and cost benefits are even more difficult to identify than on a production program. In the classic matrix, which envisions the total life cycle of the hardware, all R&D quality costs would be assigned to the prevention category, but if the R&D function is isolated as a profit center, there may be some

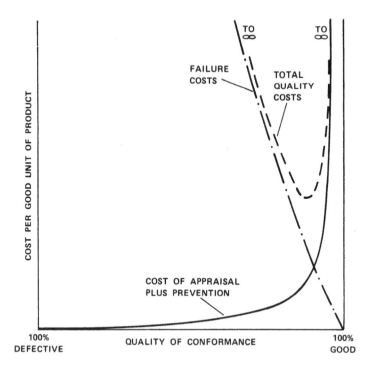

Fig. 2.2. Model for optimum quality costs. (Reprinted from J. Juran's Quality Control Handbook by permission of McGraw-Hill.)

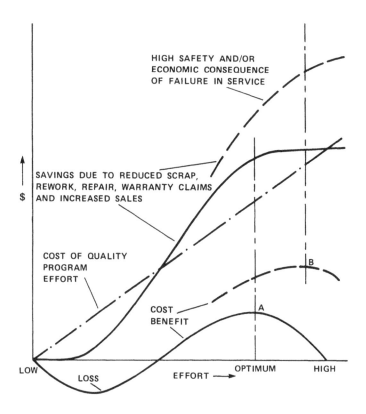

Fig. 2.3. Cost benefit vs. quality program effort. (With the permission of
    CSA this material is reproduced from CSA Z299.0-1979, Guide for
    Selecting and Implementing the CSA Z299 Quality Program Standards,
    which is copyrighted by the Canadian Standards Association, and
    copies of which may be purchased from the Association, 178 Rexdale
    Boulevard, Rexdale, Ontario M9W 1R3.)

difficulty in capturing external failures, and therefore the total picture of
quality costs would be distorted.

    It is possible to plan each project within an R&D program by assigning
specific quality levels to it, although the lowest quality level might be the one
most often employed by the research center because of cost and because the
data from these projects are used in applications involving minimum risk to
products or safety. Also, it may be a matter of basic research for which no
immediate application may be found. Since quality plans provide much more
flexibility to meet special conditions, an optimum solution may be for the research center to identify two basic levels of operation; one which is considered (for lack of a better term) "standard laboratory practice" (see Chapter
3); and the other "specified quality assurance," for which individual quality
planning is required (see Chapter 4).

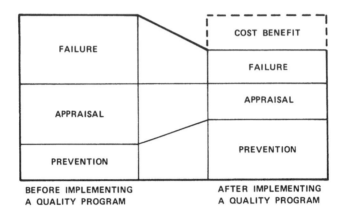

BEFORE IMPLEMENTING
A QUALITY PROGRAM

Fig. 2.4. Quality program cost benefit. (With the permission of CSA this
    material is reproduced from CSA Z299.0-1979, Guide for Selecting
    and Implementing the CSA Z299 Quality Program Standards, which is
    copyrighted by the Canadian Standards Association, and copies of
    which may be purchased from the Association, 178 Rexdale Boulevard,
    Rexdale, Ontario M9W 1R3.)

    Neill (13) describes the concepts of performing quality assessments
for each project at the Oak Ridge National Laboratory. The assessment
looks at the consequence of failure of each component or system involved in
the project in terms of: (1) human health and safety, (2) effect on the environment, (3) loss of experimental data and meeting program objectives, and
(4) effect on funding and program schedules. Oak Ridge does make exception
to projects involving studies and analytical programs where no hardware is
involved. Otherwise, each project requires this formal level of assessment.
Based on the results of the assessment and the identification of any significant quality problem areas, the project may or may not require a formal QA
Plan to be developed. The advantages to this formal assessment is that it
brings together many individuals from different fields of expertise to consider the project and its potential consequences of failure. The disadvantages
are that this type of formal assessment involves a lot of people and is, therefore, quite costly. Essentially, the formal assessment identifies each project as major or minor. The minor projects do not require a QA Plan. The
major projects have failures identified and have significant effects after which
the probability of failure is assessed. Significant quality problems must be
addressed within a quality plan developed for that project. An example of the
assessment is shown in Figures 2.5 and 2.6. The cover sheet (Figure 2.5)
for the assessment gives the background description of the project, identifies

QUALITY ASSURANCE ASSESSMENT FOR MAJOR PROJECTS		Assessment Number QAA-XYZ-65	
		☐ Draft	Date
PROJECT PHASE: ☐ Design/Construction ☐ Manufacture ☒ Operation		☒ Assessment	Date 6-1-80
Division/Program XYZ	Location Bldg. XXXX, Y-12 Plant	☐ Reassessment	Date
Project Title Coal Burning Project		REF. Engineering QAA Number Eng-QAA-1234	

1. Description (Continue on additional sheet if needed) The Coal Burning Project is a new facility for studying the coal liquifacation process which combines high temperature, high pressure, and fluidized bed operation using hydrogen rich gas for fluidization. The hydrocarbonization process can produce desulfurized char, liquid fuels, and substitute natural gas. The relative yields of these products can be controlled. The char is suitable for use with stack gas treatment in fluidized beds or traveling grate boilers. The liquid fuels include naptha, light gas oils, and heavy gas oils and contain valuable chemical feedstock. Conditions for hydrocarbonization include temperatures up to 1500°F and pressures to 150 atm. The objective of this project is to develop advanced and more efficient coal liquefaction processes and equipment. The estimated operating life for the Coal Burning Project is 10 years. The fiscal year operating cost is $700,000.

2. Assessment/Reassessment (Use worksheet on other side)

3. Assessment/Reassessment Conclusion

				Scheduled Issue Date
QA PLAN REQUIRED	☐ NO	☒ YES		July 15, 1980
QA PLAN REVISION REQUIRED	☐ NO	☐ YES	Reference QA Plan Number	Scheduled Issue Date
☐ QA PLAN REQUESTED BY MANAGEMENT				Scheduled Issue Date

4. Rationale (Continue on additional sheet if needed) In determining POF for each failure mode/concern, the assessment team took into account state-of-the-art of items and processes involved, past experience with them, & normal application of standard practices to be used by organizations involved. The assessment team recognizes the possibility that all applicable standard practices may not be fully implemented, & has considered this possibility in risk determination. Quality related standard practices selected for use on this project are provided in Attach. 1. Adequate implementation of standard practices will be assured by verification methods such as auditing, monitoring, inspection & testing. Checklists with signoffs will be used when appropriate. Procedures are reviewed & evaluated periodically. A functional responsibility chart for quality related standard practices is provided in Attach. 2. Attach. 3 gives rationale for acceptable risk determinations on the work sheet.

5. Assessment/Reassessment Meeting ☒ Minutes and list of reviewers attached

6. Approvals

Chairman of Assessment Team		Date	Division QAC		Date
	Signature	5-15-80		Signature	5-15-80
Section Head	Signature	Date 5-15-80			Date
		Date	QA Director	F. H. Neill	Date 5-15-80

7. Assessment Review

REVIEW	SCHEDULED REVIEW DATE	DATE REVIEWED	ASSESSMENT REQUIRED		APPROVAL SIGNATURES			
			NO	YES	CHAIRMAN	QAC		
1	6-1-81							
2								
3								
4								
5								

Fig. 2.5. Quality assurance assessment for major projects. (Courtesy of Union Carbide Corporation, Operators of Oak Ridge National Labs, Oak Ridge, TN.)

QA ASSESSMENT WORKSHEET
(CONTINUATION SHEET)

Assessment Number _____

Page ____ of ____

ITEM OR JOB ELEMENT	FAILURE MODE (or Concern) AND CAUSE	FAILURE CONSEQUENCE	MOF[1]	POF[2]	RISK[3]
1. Pressure Piping System	1. Leak in Pressure Piping Cause: a. Accelerated Corrosion of Pipe Wall	Delay in operation and possible safety hazard	S	U	U
	2. Leak in Hydrocarbon Vessel Cause: a. Accelerated Corrosion of Vessel Wall	Delay in operation and possible safety hazard	S	U	U
2. Data Collection System	1. Improper Operation or Total Failure of Data Collection System Causes: a. Malfunction of Data Collecting Instruments	Loss or erroneous test data	S	H	U
	b. Operator Error	Loss of test data	S	H	U
	2. Prolonged shutdown to repair XYZ Sensors Causes: a. Location of Sensors in relatively inaccessible locations	Delay in operation	S	U	U
	b. Possible high failure rate of "first-of-a-kind" sensors	Unreliable operation and potential test schedule delay	S	U	U
3. Control System	1. Loss of Control and Monitoring Causes: a. Computer Failure	Delay in operation	S	L	A
	b. Electrical Power Failure	Delay in operation	A	-	A

[1] MAGNITUDE OF FAILURE. Designate S = Significant, A = Acceptable
[2] PROBABILITY OF FAILURE. Designate H = High, L = Low, U = Unknown
[3] RISK. Designate U = Unacceptable, A = Acceptable

Fig. 2.6.　QA assessment worksheet. (Courtesy of Union Carbide Corporation, Operators of Oak Ridge National Labs, Oak Ridge, TN.)

whether it is an initial assessment or a reassessment, identifies whether a Quality Plan is required, and provides for sign-offs by the appropriate individuals. The worksheet (Figure 2.6) that is attached to this package lists the item or job elements involved and identifies the major failure modes or concerns and their causes, the consequences of failure and the magnitude and probability of failures, together with the final designation for the level of risk. Each failure mode or concern within an acceptable level of risk must have an appropriate quality action stipulated in a Quality Plan.

While Oak Ridge tends to favor projects involving hardware, Sandia Laboratories performs a "project assessment" for nonhardware projects to establish the acceptable level of risk associated with faulty outputs or project operations (14).

Formal assessments are costly and may only be desirable for high risk, high cost projects, but some sort of built-in decision process should be established to screen each project to determine if plans are required. Some laboratories have established specific Quality Plans for on-going activities [e.g., analytical labs and computer services (15)], as well as individual projects. Other labs rely on controls established in a manual section set aside for "standard laboratory practice" to govern activities and projects for which no specific QA Plan is desired.

The decision process should be spelled out in the basic administrative system for the research center. The ground rules should be clear as to which projects or activities require formal assessment, informal assessment, or no assessment and which projects or activities require specific QA Plans or may be left to operate within the constraints of "standard laboratory practice." If the rules for this decision are not clear, a representative from QA should review each activity and each new project and cause that decision to be made. If the rules are clear, QA need only audit a sample of projects periodically to assure the system is effective.

REFERENCES

1. V. Goetz, Documentation and GPM How to Reduce Paperwork, Medical Device and Diagnostic Industry, Vol. 1, No. 5 (October 1979).

2. National Aeronautics and Space Administration NHB 5300.4, Quality Program Provisions for Aeronautic and Space System Contractors, Navy Publications and Forms Center, Philadelphia, PA (1969).

3. Department of Defense MIL-Q-9858A, Quality Program Requirements, Government Printing Office, Washington, D.C. (1969).

4. United States, Code of Federal Regulations 10 CFR 50, Appendix B, Quality Assurance Criteria for Nuclear Power Plants and Fuel Reprocessing Plants, Government Printing Office, Washington, D.C. (1970).

5. American National Standards Institute, ANSI/ASME NQA-1-1979, Quality Assurance Program Requirements for Nuclear Power Plants, New York (1979), American Society of Mechanical Engineers.

6. A. Feigenbaum, <u>Total Quality Control</u>, McGraw-Hill, New York (1974).

7. J. Juran, <u>Quality Control Handbook</u>, McGraw-Hill, New York (1974).

8. American Society for Quality Control, Matrix of Nuclear Quality Assurance Program Requirements, Milwaukee, Wisconsin (1976).

9. Dr. J. S. Foster, Quality and Productivity, 35th ASQC Quality Congress and Exposition, San Francisco (1981).

10. G. Roberts, Quality Assurance in R&D, <u>Mechanical Engineering</u>, Vol. 100, No. 9, p. 41 (1978).

11. Atomic Energy Commission, Division of Reactor Development and Technology RDT F2-2, Quality Assurance Program Requirements (1973).

12. Canadian Standards Association CSA Z299-0-1979, Guide for Selecting and Implementing the CSA Z299 Quality Program Standards, Rexdale, Ontario (1979).

13. F. Neill, Quality Assurance Programs in Research and Development, Quality Assurance in a Large Research and Development Laboratory, ASQC Seventh Annual National Energy Div. Trans., Houston (1980).

14. J. Calek, Applications of Quality Assurance/Quality Control Concepts to Non-hardware Project Management, SAND 79-1921, Albuquerque (1979).

15. Private communication to the author.

16. Reprinted from S. Marash, Quality Assurance Systems Requirements for Nuclear Power Plants — Part 1, <u>Journal of Quality Technology</u>, Vol. 5, No. 3 (July 1973), Milwaukee.

Chapter 3

STANDARD LABORATORY PRACTICE

I. DEFINITION

"Standard laboratory practices," as used in this book, comprise the basic quality level for operating a research laboratory. In this situation, the verification and checking is at a minimum and left to the discretion of the research project leader. Management may elect to review the details of a research report on a case-by-case basis. The evidence of quality is within the report which is prepared on completion of the project (for longer projects, detailed interim reports may be desirable).

The basic quality level may be appropriate for:

1. Low risk projects with a low potential of liability for the data, data used for preliminary information, or projects without a direct application.
2. Studies and literature searches.
3. Consulting or routine field service work.
4. Low risk projects using low cost test apparatus.
5. Projects which are intended to be repeated with adequate checking and documentation later.
6. Standard testing using specific accepted industrial test procedures which include adequate provisions for control and documentation.

As the majority of projects within a research center tend to move toward requiring higher risks (financial or safety), the basic quality level should be modified to incorporate the QA systems described in other chapters of this text.

The one exception is calibration of test equipment. The principles of control of measuring and test equipment presented in Chapter 8 should be considered basic to any definition of "standard laboratory practice." Because of its importance and for general clarity, this subject has been assigned a separate chapter. But it is not intended that a "special" level of quality assurance should have to be assigned before a research center takes strong measures to ensure the quality of its measurements. The reputation of a research organization rests too heavily on the data it acquires to leave its calibration system to individual perceptions of what is necessary. An organized, controlled system is essential.

Compliance with the basic quality level should be verified by management by periodic audits of projects selected at random (see Chapter 12, Section II. B).

II. ADMINISTRATIVE PROCEDURES

Each laboratory and each section within the laboratory has a different function to perform and has often developed its own means of accomplishing those functions. Since the tasks performed by technical sections vary day-to-day, there is an understandable reluctance to document rigid operating methods that may not be universally appropriate for all circumstances. Nevertheless, there is a need to establish certain basic standards within each research section which allow the dissemination of section and laboratory management policy to the persons working in that particular section. While the research center can and should establish minimum criteria for the work standards to be followed by each section, it falls upon the individual section manager to develop appropriate procedures to interpret the laboratory policy as it applies to that section, and to specify the means of implementation of laboratory policy within the framework of the responsibilities of section personnel. Section managers have the option to add to the requirements of the overall center's standards.

Once standards are developed, it is obviously appropriate for them to be made available to each person within the section. Standards serve no purpose whatsoever gathering dust on a supervisor's bookshelf while employees are having to learn by word of mouth. When procedures are out of sight they are out of mind.

One of the areas for consideration for standard procedures should be the identification and assignment of responsibilities within the section for assigning and performing work. If the technical section is normally involved in providing routine chemical analyses which require very little detailed advanced planning, the assignment and performance of that work would be more decentralized than a major development project which must be designed, inspected and built for a long program of testing.

Additional measures to be considered are those involving the protection of proprietary and classified material. Procedures should be developed to clearly indicate responsibilities of personnel with respect to clearing technical papers and speeches or news releases with a specific individual or coordinating committee. This may include a review by individuals with legal or patent responsibility. If the research is performed in accordance with a government contract for which security requirements are imposed, procedures would have to be in place to meet the government control agency's requirements for declassifying of documents and retention and disposal systems for classified materials. Finally, the procedures should specify requirements for planning, documenting, independently reviewing, and reporting of the research work. Detailed discussion of these areas follows.

III. PLANNING

A key element in effective (productive) R&D is planning. Veronda stated in his article, Optimizing Administrative Controls, that:

> In looking at many examples of the most outstanding and original work including many Nobel prizes, the planning and organization of even the purest research is very important and does not hamper activity if the person doing the work establishes the plans and milestones (1).

The planning process for research projects should be well thought out in advance because of the potential impact to a company's technical and financial future and because the cost of R&D dictates the need to get the most out of every project. Some companies go through an extended planning cycle using supporting technical committees and corporate management reviews. An example of this process is given in Appendix A.

Specific plans for each project should be developed and include milestones and budget estimates for completion. The example shown in Figure 1 provides spaces for a project description and scheduling. For a basic quality level project, there may be no QA specifications entered (a standard QA plan could be developed once and referenced for all such projects).

To reiterate the statements made in Chapter 2, the ground rules for selecting the basic quality level over one individually planned should be made abundantly clear to everyone. Otherwise, assistance should be sought from the QA organization to review each project prior to organizing the plan or starting work.

IV. RESEARCH DOCUMENTATION

Documentation of R&D work is important regardless of the level of the quality assurance perceived to be appropriate. Any information, memos, meeting minutes, instrumentation calibration curves, or special references that have a bearing on the direction the project has taken, the data acquired, or the interpretation of results should be included in the project files. There are, however, some key areas that warrant special discussion.

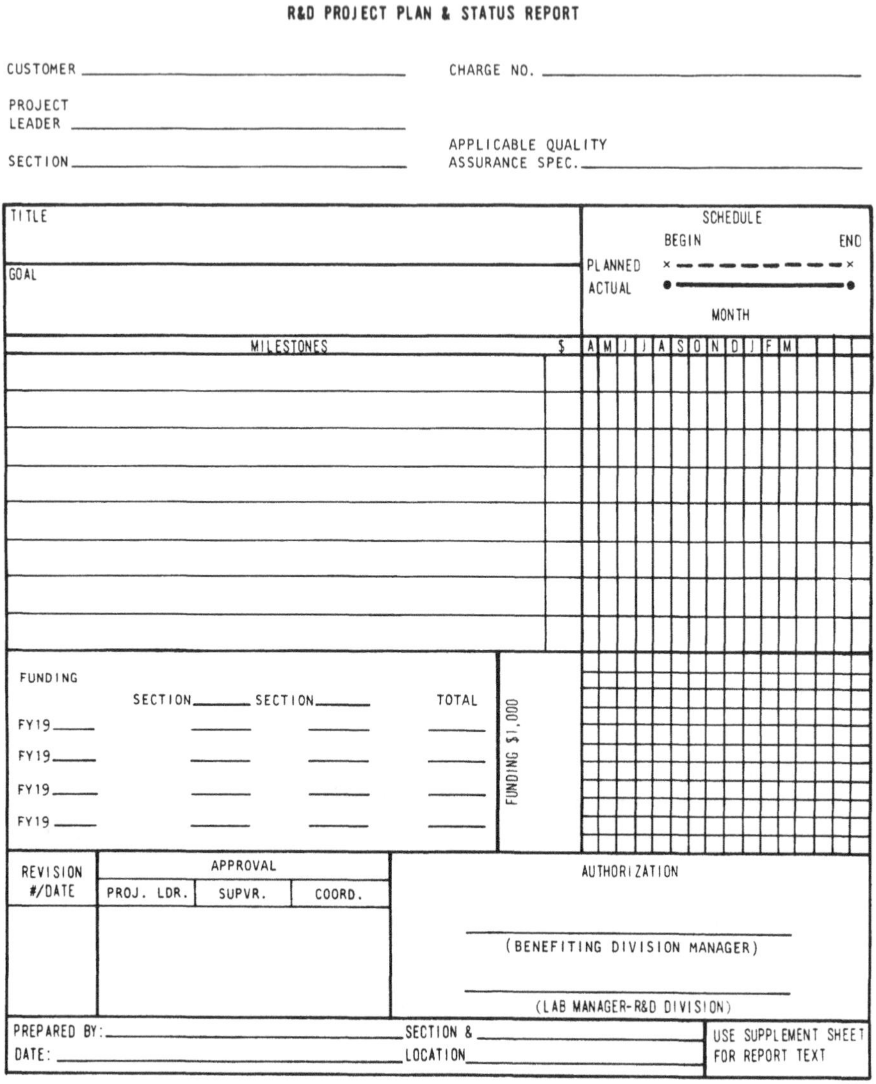

R&D PROJECT PLAN & STATUS REPORT

CUSTOMER ______ CHARGE NO. ______

PROJECT LEADER ______

SECTION ______ APPLICABLE QUALITY ASSURANCE SPEC. ______

TITLE

GOAL

SCHEDULE

BEGIN END

PLANNED × — — — — — — ×

ACTUAL ● ———————— ●

MONTH

MILESTONES $ A M J J A S O N D J F M

FUNDING

SECTION ______ SECTION ______ TOTAL

FY19 ______

FY19 ______

FY19 ______

FY19 ______

FUNDING $1,000

REVISION #/DATE

APPROVAL

PROJ. LDR. SUPVR. COORD.

AUTHORIZATION

(BENEFITING DIVISION MANAGER)

(LAB MANAGER-R&D DIVISION)

PREPARED BY: ______ SECTION & ______

DATE: ______ LOCATION ______

USE SUPPLEMENT SHEET FOR REPORT TEXT

Fig. 3.1. R&D project report of milestone and budget performance. (Courtesy of Babcock & Wilcox.)

A. Work Orders and Customer Communications

An area that is extremely sensitive to misunderstanding and contention is
that involving what the researcher is supposed to do, the expected schedule
for the task, and the anticipated cost. It would seem that if the proper
planning were done in the beginning using a well documented process such
as that shown in Figure 3.1, then few problems should come up regarding
what the customer wants and gets. Unfortunately, this is not always the
case. For research to support an internal company design function and
where formal contracts are not used, there is a tendency towards "gentle-
men's agreements" to get the work out on time. When costs begin to
mount, usually due to undocumented changes, the design organization may
have second thoughts about whether the data was really needed and may
request the research group to stop where they are and send whatever data
they have verbally. The research group may be caught in the middle of
data acquisition where shutting down the apparatus will cost not much less
than proceeding to the planned conclusion of the test. To the researchers,
the complete test is necessary to find answers applicable to a broad range
of problems applying to more than one division. And no research organization will stand behind information transmitted only verbally — there must
be a written report. The sponsoring division has no additional funds, is
not concerned about data affecting other divisions — let them fund their own
tests — and does not need paper, just numbers. Besides, there is a sus-
picion that the R&D group was riding their charge number a little too heavily because it was a "live" number for a panic job. So much for "gentle-
men."
       The only protection against this kind of scenario is written communi-
cations maintained with the project files. Whenever changes are discussed
and agreed upon, the project leader should document the understanding together with the anticipated cost and schedule impact and send a copy to the
sponsoring organization. Updating project authorizations with consecutive
revision levels (Figure 3.1) and obtaining sign-offs will assure all parties
are in accord and reduce the chances of misunderstanding and conflict.

B. Purchasing Documents

Of course, the Purchasing Department has copies of purchase orders for
your project — what was the PO number again? Purchasing serves many
organizations and projects within the center, and their needs for data retrieval are different from a project leader's needs. For quick reference, a
copy of all purchasing documentation should be kept in the project files for
the same reasons given in Section IV.A above. The technical specifications,
costs, and schedules should be spelled out carefully so the supplier (and
buyer) will not misunderstand what is expected. Any changes should be documented and approved by all parties, not just the buyer. If certificates are

nonstandard procedures. Sometimes the whole purpose of the test is to
required, they should be spelled out as to what <u>kind</u> of certification (certificates of conformance are different from material test reports) and to what
standards or specifications including dates or editions. It is worth the
price of a telephone call to make sure the supplier understands before the
work is begun or shipment is made.

C. Test Article Configuration

It is important to know exactly what was tested. Many factors can introduce
variability into test results and, unless the article itself has been adequately
described, any efforts to duplicate the tests at a later date would be frustra-
ted. Hoe (2) distinguishes between the test facility and the test article (section) — the facility being the "system, structures, or equipment necessary
to provide the test conditions during the operation of the test." The test
section is the device being tested and may be the production item itself, a
scale model of the device, a mockup or simulation of the device, or a test
specimen as in the case of materials. Any of the characteristics of either
the test facility or the test article that have a bearing on the validity of the
calculations or conclusions of the test should be well documented. Detailed
drawings from the group responsible for designing the test article are the
best means of establishing the test article configuration <u>provided</u> the drawings reflect as-built dimensions. Alternately, separate documentation
should verify that the drawing dimensions were faithfully followed during
construction of the test article and its installation into the test facility.

D. Laboratory Notebooks

An indication of the importance of the laboratory notebook is shown by
Hughes and Ennis in their statement that, "It is the function of the Research
Administration to ensure that the guidance provided by the legal department
on the maintenance of laboratory notebooks, particularly the recording of
inventions, is thoroughly disseminated to all concerned" (3).

Standards for making entries and revisions to the notebooks should be
spelled out. The usual accepted practice is for the entries to be made leg-
ibly in ink, with no erasures permitted (see Figure 3.2). Any changes or
corrections to the entries should be made by drawing a single line through
the erroneous entry and having the correction entered adjacent to the error,
dated, and initialed. Some organizations require that the laboratory entries
be witnessed and initialed at the end of each day. Although bound notebooks
do not absolutely require witnessing initials to be placed at the bottom of
each page, there is an advantage to this practice. Individual pages could be
brought into court during a patent action without disclosing the remaining
notebook. This would protect additional proprietary information for the
company (4). Instructions should be included as to the type of content to be
included in the laboratory notebooks; for example, the identification of specific measuring and test equipment being used, backup literature being referenced, basic assumptions and objectives of the test, and appropriate re-

TITLE SENSOR QUALIFICATION TEST   Project No. 2043
                                     Book No. 1   84

From Page No.___

> Today the Autoclave will be taken to the
> hydro condition of 3125 psig @ 150°F.
>
> Before the pressure could be increased the springs
> on the Shaker had to be adjusted to counter-act
> the pressure load on the shaft.
>
> Below is the calculation used to determine the
> required spring compression.

0900 HRS   810 lb/inch × 4 springs = 3240 lb/in.

               3125 psi × .785 in²   = 2453 lbs load on shaft
                          shaft area

           therefore $\dfrac{2453\ lbs}{3240\ lbs/in}$ = .7571 inch spring
                                                            compression req'd.

> The spring compression was released and a
> reference measurement taken. The safety shim
> was installed and the springs compressed
> 0.75 inches and the lock collars positioned
> to transfer the spring load to the shaft.
>   2670 JJK 1-15-82

1300 HRS   1196.0 E.T.   150.1°F   Pressure 2570 psig
1312 HRS   150.6°F   3125 psig
             No leaks observed

1330 HRS   @ 3125 psig load Cell output on shaker
            - 123 lbs. This load will be offset
            with the shaker during dynamic tests.
            This small load indicates spring
            compression is good.
1340 HRS   START RECORDING DYNAMIC DATA
1445 HRS   Electrical Checks performed.
                                        To Page No.____

Witnessed & Understood by me,	Date	Invented by	Date
G.W. Raman	1/15/82	Recorded by J.Jeffrey Kidwell	1-15-82

Fig. 3.2. Laboratory notebook. (Courtesy of Babcock & Wilcox.)

marks at the end of the project as to whether those objectives were met.

The laboratory notebooks should include or reference the procedures used by number and revision. Many of these procedures have been proven by extensive round robin testing by several laboratories and are published in the ASTM Standards (5), American Chemical Society Specifications for Reagent Chemicals (6), and the United States Pharmacopeia (7) to name a few. The procedures might only need to be referenced in the project files. In other cases, however, there is a need for preparation of new standard or

develop a procedure. If there is not a document serializing system that
permits referencing of the procedure number and revision in the lab notebooks, copies of the procedures should be included in the project files.

E. Reports

A formal reporting system should be used to permanently document the results of the R&D work. This reporting system should include quarterly
progress reporting against milestones and expenditures, as well as a system for generating final reports at the conclusion of the research project.
        Financial control measures work hand in hand with technical controls.
An indication of over or under expenditures, or an excessively long completion time could be the result of either an inadequate or an excessive quality
verification system. Monitoring of project performance should include periodic review of schedule and cost information. Monthly or quarterly reporting (see Figure 3.1) of milestone completions and budget expenditures is appropriate, with a requirement for the project leader to highlight unusual
variances (e.g., plus or minus 5% of budget or X dollars, whichever is
greater).
        Specific recommendations for report content are given in Chapter 11,
but as a rule the higher the management level the report is intended to
reach, the more it must be screened to cull out unnecessary detail and to
provide concise summaries highlighting key issues. Reports should be directed at the level which is responsible for approving the budget. Recommendations should be included to recover from undesirable trends. Since
communication is a two-way process, management's reactions in the form
of specific approvals of redirection should be documented in the project
files. To repeat, if a change in workscope or budget is authorized, project
tracking sheets should clearly reflect this, with approvals to at least the
same level approving the initial budget and work scope. Standards for this
action should be clearly stated in writing.
        Finally, there should be some system of identification and filing for
R&D work to permit information retrieval at a future date. The use of
some unique serial numbering system is recommended.

V. REVIEW OF RESULTS

Encouraging technical publication with its associated peer review may be an
effective means for ensuring appropriate care in taking and interpreting
data. But for an industrial research center, only a small percentage of research will find its way into general publication. Most of the information
is factored into designs or processes that are proprietary. Therefore,
other steps must be considered.

For critical applications, the options of a formal design review or a detailed independent technical review as presented in Chapter 2 should be considered. At the minimum, an engineering manager should expect that someone other than the principal investigator is aware of the data being transmitted from his organization. Varying levels of management approval are usually required for different types of reports. Memo reports on the interim status of a project should be at least informally cleared with the group supervisor. If the supervisor is on the list of copy recipients, his initials on the original is a quick and simple indicator of his awareness of the information being released. More formal reports at the conclusion of a project (or a major phase of it) would be subject to successively higher levels of management approval.

As in the case of patents, the witnessing initials at the bottom of the lab notebook may also be a review process, but this should be spelled out beforehand. Review and approval is substantially different from "witnessed and understood."

APPENDIX: ANNUAL PLANNING CYCLE FOR A CORPORATE R&D CENTER

Research and development (R&D) plans are prepared and administered to support the comprehensive business plans of operating units or divisions of the corporation. The broad objective of an R&D plan is to provide a technical base to competitively support the operating unit's or division's business position and to discipline the continuing development of new technical knowledge, products, and services for future growth and new ventures. The primary responsibility for preparing and executing an R&D plan is assigned to the operating divisions. In this responsibility is included the initial identification of R&D needs and subsequent preparation, evaluation, and execution of an approved plan. At the same time, the R&D division maintains a continuing cognizance of the operating division's R&D activities, and should evaluate and concur with all operating division R&D plans. The primary basis for such concurrence is the supporting relationship of the R&D plan to the operating division's comprehensive business plan.

The parallel roles assigned to the R&D and operating divisions are interrelated through the operation of an R&D steering committee for each operating division. It is the responsibility of the steering committee to implement and monitor the R&D planning process for its respective operating division. A typical steering committee may include the following representatives:

Operating Division Coordinator
Operating Division Technical Manager
Operating Division Planner
R&D Division Coordinator

R&D Division Technical Manager
Corporate Operational Planning Representative
Major Project Leaders

This steering committee must evaluate and agree on the R&D planning schedule, R&D priorities, program objectives, scope, schedules, and budgets,
and the R&D plan summary, which is contained in the comprehensive business plan for the R&D division. R&D planning is a continuing process, documented on an annual basis and scheduled to parallel the preparation of the
comprehensive business plans. The R&D plan summary, which is an integ-
ral part of the business plan, is prepared early in the planning cycle and
incorporated with the other components of the business plan. A final detailed R&D plan containing this plan summary and including approved project
or program descriptions (see Figure 3.3) is issued prior to the beginning of
the plan year.
     The R&D planning process takes place in several phases during the
year preceding the year of implementation. The following process is illustrated using a calendar year cycle.

A. Preliminary Planning (January-March)

The planning cycle starts with the development of the expected business environment, critical issues and strategies by the operating division management and planning staff. Having developed tentative goals and strategies for
the division, the operating division and R&D division coordinators prepare a
schedule for the R&D plan preparation consistent with that for the comprehensive business plan. Operating division project and program leaders com-
pile proposed project activities and objectives. Joint planning discussions
are held between operating division and research and development project
leaders and program leaders to review, evaluate and set priorities for proposed project activities. Initial steering committee meetings are then
called to determine technical needs and review technical objectives. Information from external sources such as marketing, planning, manufacturing
and finance is considered. Proposed activities are then evaluated by the
steering committee and the priorities confirmed or revised on the basis of
available information. Preliminary program budgets are also determined
and balanced against known financial constraints. Necessary adjustments
are made through interchange among the technical staff, steering committee
and other management. These are based on technical need, risk analysis,
financial constraint and priority. The preliminary planning process is cri-
tically important and serves as the basic foundation for the final R&D plan.

R&D PROJECT PLAN

CUSTOMER _____   CHARGE NO. _____

PROJECT
LEADER _____

                                             APPLICABLE QUALITY
SECTION _____   ASSURANCE SPEC. _____

PROGRAM TITLE
PROJECT OR WORK PHASE TITLE
BACKGROUND, STATUS & JUSTIFICATION (CITE BENEFITS & EXPECTED RESULTS)*
SUMMARY OF PLANNED ACTIVITY FOR THE LIFE OF THE PROJECT* (EMPHASIZE CURRENT YEAR WITH A DISCUSSION OF MILESTONES)

Fig. 3.3. R&D project description and justification. (Courtesy of Babcock
& Wilcox.)

B. R&D Plan Summary Preparation (April-June)

The operating division prepares an R&D plan summary in accordance with
the recommendations of the steering committee. This plan receives a
preliminary review by operating division and R&D division management.
Discrepancies or topics which cannot be mutually agreed upon are identified. The R&D division issues a letter which outlines the status of agreement between operating division and R&D management. The operating division then incorporates the R&D plan summary into its comprehensive business plan with the changes they feel are appropriate and submits the total
plan to corporate operational planning.

C. Program/Project Resolution (July-August)

Program and project budgets, schedules, work scopes or priorities which
are disputed or not yet clearly defined are negotiated and resolved by the
operating division and R&D program and project leaders. When program
and project leaders are not able to reach agreement, they seek assistance
from their management.

D. Comprehensive Business Plan Review (September-October)

The R&D division reviews the completed comprehensive business plans for
each operating division and makes an assessment of the adequacy of the
R&D plan summary with respect to the business plan. Results of the review
are communicated to the operating division head. Agreement must be established among the operating division, the R&D division and corporate operational planning before completing the final detailed R&D plan.

E. Preparation of Final R&D Plan (November-December)

Project authorization and program summary forms (Figure 3.1) are prepared by the project leader and processed for approval. Revisions and modifications are incorporated into an updated summary R&D plan. The operating division and research and development division coordinators correlate
the necessary program and project forms with the summary R&D plan and
prepare the complete plan. The operating division coordinator issues the
final plan by January 15.

F. Implementation and Execution (January 15 through Plan Year)

Accounting forms are prepared for each project. They are approved by
project leaders, management and coordinators. Work begins on approved
projects. Modifications of the plan are identified and incorporated throughout the plan year.

REFERENCES

1. C. Veronda, Optimizing Administrative Controls, In Improving Effec-
    tiveness in R&D, Edited by R. Cole, Thompson Book Company, Inc.,
    Washington, D.C. (1967).
2. R. Hoe, Managing the Engineering and Research Laboratories, In Man-
    aging Engineering and Research, Edited by D. Karger and R. Murdick,
    Industrial Press, New York (1963).
3. G. Hughes and H. Ennis, Jr., Department of Research Administration,
    In Handbook of Industrial Research Management, Edited by C. Keyel,
    Reinhold Book Corporation, New York (1968).
4. Managing Engineering and Research, Edited by D. Karger and R. Mur-
    dick, Industrial Press, New York (1963).
5. Annual Book of ASTM Standards, American Society for Testing and Ma-
    terials, Philadelphia, PA.
6. Reagent Chemicals, American Chemical Society Specifications, Edited
    and Produced by American Chemical Society Publications, 4th Ed.,
    Washington, D.C. (1968).
7. The United States Pharmacopeia USP XX, The National Formulary NF
    XV, United States Pharmacopeial Convention, Inc., Rockville, MD
    (1980).

Chapter 4

PROJECT QUALITY
ASSURANCE PLANNING

I. PROJECT TECHNICAL PLANS

As work requests are received by the R&D organization, a project leader should be identified to carry out the requirements of the research function. This project leader has the primary responsibility for establishing the technical standards <u>and</u> for implementing quality assurance requirements for all projects. He has the overall responsibility for the technical performance of the project, and is in the best position to schedule the required checking, inspection and documentation functions (1). The project leader and a representative from the QA organization should evaluate work requests for research activities and determine, in light of customer requirements and the nature of the work, which of the research center's QA disciplines are required for that specific project.

The experimental concept for a project can be described in a project technical plan which, together with a QA plan will establish the overall requirements of the project. If the research is performed for an outside customer, the project technical plan may be replaced by a statement of work as a part of the contract proposal. In either case, project technical plan or a contract statement of work, the elements to be included in that planning function should be similar.

A suggested list of topics to be covered by the project technical plan is:

1. Introduction with purpose of the project and background
2. Description of the activity; that is:
    a. Type of experiment or analysis
    b. Philosophy, basic principles and limitations involved in the choice of the test or experiment design concept
    c. Parameters to be investigated
3. Test apparatus
    a. Description of how the test section or experiment is to be constructed

    b. Identification of materials, parts or components having an effect on
        the results
    c. Identification of measurements having an effect on the results
    d. General identification of special processes, e.g., welding, heat
        treating, brazing, NDE to be used which may affect the test results
4. Test program: Description of the overall test program and identifica-
    tion of specific test procedures to be used or developed to implement
    the overall program
5. Description of data acquisition and reduction methods
6. Description of data evaluation methods

    Because the project technical plan is used as a formal means of
agreement between the customer and the research center, the approval and
revision control of the plan also should be formalized.

II. PROJECT QA PLANS

The QA plan is developed in conjunction with the project technical plan to
selectively impose verification activities and processing controls. QA plans
may be narrative or of the checklist variety.
    Figure 4.1 illustrates a plan using a checklist format with specific im-
plementing procedures noted adjacent to each of the applicable QA elements
to be imposed on the project. The implementing procedures provide the
specific detailed activities to be followed when performing the quality assurance actions required for this project. Additional short notes can be entered
in the Remarks column and supplemental sheets are only added when necessary for additional clarification or to make special provisions not otherwise
covered in the implementing procedures. This format has the advantage of
rapid preparation and, where implementing procedures are used, allows the
project leader to know immediately which procedures he should have on hand
for that project. The final page of this plan includes a checklist of all the
quality records expected to be generated for this project. The checklist can
be used by the project leader to verify that his files are complete. This illustration implies that the user must have a QA manual, a QA plan, and detailed implementing procedures. It assumes the QA manual is a policy manual and does not give sufficient direction to permit detailed implementation.
    Figure 4.2 illustrates a variation of the checklist QA plan. Most of
the detailed instruction is contained within the quality assurance manual.
The QA plan indicates which sections of the manual apply and an additional
column is provided that permits the indication of areas where amplification
is required. Figure 4.3 is the supplemental sheet with clarifying entries
corresponding to each of the check marks in the W/C column of the QA plan
index. Since no implementing procedures are referenced, assumption is
that the user need only have the QA manual and the QA plan to understand
what was expected of him during the course of the project.

QUALITY ASSURANCE PLAN

CUSTOMER NPGD, Lynchburg   QA PROJECT NO. 84526

CUSTOMER CONTRACT NO. 792-0416-51   REVISION ___0___ DATE 11/16/81

R&DD PROJECT NO. E 97458

PROJECT Fuel Assembly Impact Test   DATE 8/12/81

IN ACCORDANCE WITH CUSTOMER SPECIFICATION 04-1653-00   DATED 7/8/80

PREPARED BY: APPROVED BY:
QA ADMINISTRATOR S. L. Molnar 11/17/81   PROJECT LEADER R. B. Smith 11/18/81

APPROVED BY: APPROVED BY:
QA MANAGER G. W. Roberts 11/17/81   SUPERVISOR H. M. Jones 11/17/81

THE SECTIONS OF THE R&D QUALITY ASSURANCE MANUAL DESIGNATED BELOW ☒ (AND IMPLEMENTING PROCEDURES REFERENCED IN THE MANUAL) ARE APPLICABLE TO THIS PROJECT.

SECTION		REMARKS
1.0 INTRODUCTION	☒	
2.0 QA PROGRAM	☒	MANUAL REVISION 6/20/80
3.0 DESIGN CONTROL		
DESIGN REVIEW	☒	11005-01 dated 2/15/80
INDEPENDENT TECHNICAL REVIEW	☒	11027-03 dated 11/20/80
CALCULATIONS	☒	11014-00 dated 4/18/79
COMP. PROGRAMS	☒	11006-03 dated 5/12/80
4.0 PROCUREMENT DOCUMENT CONTROL (QA REVIEW)	☒	16000-04 dated 12/4/80
5.0 INSTRUCTIONS, PROCEDURES & DRAWINGS		
DRAWINGS	☒	19001-05 dated 3/27/81
ROUTE SHEETS	☐	
INSPECTION CHECKLISTS	☒	19022-03 dated 8/15/80
ADMIN. PROCEDURES	☒	12005-02 dated 3/12/79
TECHNICAL PROCEDURES	☒	15009-04 dated 7/26/80
6.0 DOCUMENT CONTROL		
ADMIN. PROCEDURES	☒	12005-02 dated 3/12/79
DRAWINGS	☒	See Section 5
INSPECTION CHECKLISTS	☒	See Sections 7 and 10
PROPOSAL	☐	
PROJECT TECHNICAL PLAN	☒	17004-02 dated 2/26/79
QA MANUAL	☒	
QA PLANS	☒	17002-03 dated 3/15/80
ROUTE SHEETS	☐	
FINAL REPORT	☒	20401-01 dated 4/28/79 - An independent technical review of the program shall be performed and documented prior to issuance of final report.
TECHNICAL PROCEDURES	☒	See Section 5
RELEASE OF DATA	☒	18005-03 dated 12/16/79

Page 1 of 3

Fig. 4.1. Checklist style quality assurance plan. (Courtesy of Babcock & Wilcox.)

QAP No. __XYZ-1_____ Rev. __0____ Page 2 of __6____

Project: __Rock Characterization and Analysis_____

QA PLAN INDEX

	Applies	W/C[a]	REV[b]
1.1 Organization	●	O	
1.1.1 QAO Organization	●	O	
1.1.2 QAO Responsibilities	●	O	
2.1 QA Program	●	O	
2.1.1 Applicability	●	●	
2.1.2 QA Plans	●	O	
2.1.3 Personnel Qualifications and Training	●	O	
2.1.4 QA Program Review	●	O	
2.2 QA Planning	●	O	
2.2.1 Project QA Plan	●	O	
2.2.2 Activity QA Plan	O	O	
2.2.3 Revision	●	O	
2.2.4 Close-Out	●	O	
2.3 Job Quality Plan—applies to work performed by Facilities Engineering Section and Major Projects and Planning Organizations		O	
2.3.1 Requirements		O	
2.3.2 Preparation and Approval		O	
2.3.3 Revisions		O	
2.3.4 Records		O	
3.1 Design Control and Method Review	O	O	
3.1.1 Design Planning	O	O	
3.1.2 Project Approach and Method	●	●	
3.1.3 Design and Engineering Control	O	O	
3.1.3.1 Design Document Control	O	O	
3.1.3.1(3) Non-Official Engineering Drawings	●	O	
3.1.3.2 Design Criteria	O	O	
3.1.3.3 Design Configuration Control	O	O	
3.1.3.4 Engineering Hold Points	O	O	
3.1.3.5 Modifications	O	O	
3.1.4 Calculation (Data) verification	O	O	
3.1.5 Interface Controls	O	O	
3.1.6 Design Review and Verification	O	O	
3.1.6.1 Design Data on OED	O	O	
3.1.6.2 Engineering Work in Support of Design	O	O	
3.1.6.3 Formal Design Review	O	O	
3.1.6.4 Development Testing	O	O	
3.1.7 Revisions	●	O	
4.1 Procurement—Applies if a purchase requisition is processed (procurement or subcontract)		●	
4.1.1 Procurement/Subcontract Document Preparation and Approval		O	
4.1.2 Source Inspection Activities		O	
4.1.3 Supplier/Subcontractor Selection		O	
4.1.4 Procurement Document Close-Out		O	
5.1 Instructions, Procedures, and Drawings	O	O	
5.1.1 Ident. & Content	●	●	
5.1.2 Transmit of QA Req.	O	O	
5.1.3 Compliance	●	●	

(a)W/C—With clarification presented in this QA Plan
(b)REV—Indicates revision number in which changes were made

Fig. 4.2. QA plan index. (Courtesy of Battelle, Pacific Northwest Laboratories, Richland, WA.)

2. QA PROGRAM

    2.1.1 The analytical lab and computer services shall each have an Acti-
    vity QA Plan in accordance with the Quality Assurance Manual.

3. DESIGN CONTROL AND METHOD REVIEW

    3.1.2 The project approach and method is documented in the Project Plan
    Document. Review and approval by the Project Manager and the
    (appropriate) Section Manager shall be noted by signature and
    date on the cover page or other means traceable to the Project
    Plan Document.

In addition:

    Calculations and analyses performed in support of project results
    and reported to the sponsor shall be documented and traceable from
    the primary data and its source, through the assumptions and/or
    interpretations made, to the corresponding calculation and/or re-
    ported analysis.

4. PROCUREMENT

    4.1 The applicable parts of Sections 9.1 and 12.1 shall apply if re-
    ceiving inspection is to be applied to a specific procurement.

5. INSTRUCTIONS, PROCEDURES, AND DRAWINGS

    5.1.1 Written procedures are required for collecting, identifying, packag-
    ing, and shipping field samples.

    5.1.3 This subsection applies to the procedures listed in 5.1.1 above.

In addition:

    A description of the research procedures and methods used (or de-
    veloped) shall be recorded in the Laboratory Record Book. (Note:
    It is only necessary to make reference to established procedures
    available in the existing literature.)

6. DOCUMENT CONTROL

    6.1 This section shall be applied to the procedures identified in 5.1.1
    above. Additional information is given below:

Fig. 4.3. Narrative supplement to QA plan index. (Courtesy of Battelle,
Pacific Northwest Laboratories, Richland, WA.)

The advantage of the first system shown is that changes may be made
to detailed implementing procedures without changing the manual, and a
manual can be provided to prospective customers without giving all the detailed procedures wide distribution outside the company. This helps protect
proprietary information. The disadvantage of the first system is that there
is more overall paperwork needed by the project leader to understand the
QA requirements. The second system shown in Figures 4.2 and 4.3 reme-
dies the problem. The disadvantage to the second system is that any detailed
changes to the system must be factored into the quality manual and are

immediately known to all manual holders. This may require obtaining a customer review and approval of the overall manual for a minor change in implementation technique.

A third alternative to QA planning is to use a narrative form illustrated
by Figure 4.4 and Table 4.1. Figure 4.4 shows the basic QA plan with a cover sheet for appropriate signatures and a narrative section describing the
project facility description, the organization and responsibilities for the project direction, and an introduction to the quality assurance actions to be taken
for the project. Table 4.1 is the table of special QA actions that were identified to handle the concerns generated during the QA assessment (see Chapter 2). This table provides very specific direction as to the QA actions to be
taken to compensate for or to prevent a failure or degradation of quality.
This system of planning used together with the quality assessment provides
a high level of specialized control over the project activities without having
to reference any other documentation. This plan could be written as an in-
tegral part of a project technical plan to provide a comprehensive technical
description of the project and its quality assurance actions.

As the research project grows larger, more elements are identified as
being critical to the data. The design and construction of a large high energy
physics test apparatus, such as Brookhaven's ISABELLE or Princeton Plas-
ma Physics Laboratory's TOKAMAK Fusion Test Reactor, are examples of
projects warranting a full program status with full-time QA personnel assigned. These QA programs would appear very similar to a traditional QA
program for any high technology product line. Verification activities, however, should not be allowed to concentrate only on measuring the physical
aspects of the test apparatus to assure compliance with the design. Verification should include the conceptual design and the data resulting from the
experimental use of the apparatus.

III. IMPLEMENTATION

Once the project technical plan and QA plan are developed and agreed to by
the research center and its operating division or customer, the implementation of these requirements is primarily the responsibility of the project
leader. The project leader assures that the design, procurement, construction, test, and checkout activities are conducted in accordance with the needs
of the research project as defined in the experimental concept. Once the
construction is complete, the project leader may take over the operation of
the test facility and perform whatever testing and evaluation is called for by
the technical plan. The research center's independent QA function evaluates
the project leader's actions to assure compliance with the requirements defined in the QA plan. Through the use of surveillance and formal audits, the
QA organization is then in the position at project completion to certify that
the research results were obtained in accordance with the project technical
and QA plans.

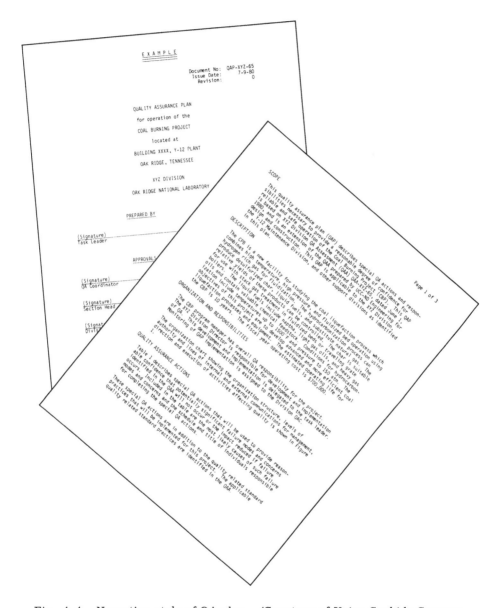

Fig. 4.4. Narrative style of QA plan. (Courtesy of Union Carbide Corporation, Operators of Oak Ridge National Labs, Oak Ridge, TN.)

Table 4.1 Special QA actions

Potential significant failure modes or concerns and causes	Special QA actions	Schedule	Responsibility
1. Leaks in pressure piping system Cause: Accelerated corrosion of pipe wall	Conduct pressure test and measure wall thickness of piping on a periodic basis. Prepare procedure.[a]	Complete special procedure 7-1-80	XYZ Task Leader[b] and QA&I Inspector
2. Leaks in hydrocarbonizer vessel Cause: Accelerated corrosion of vessel wall	Pressure test vessel and visually inspect interior of vessel on a periodic basis. Prepare procedure.[a]	Complete special procedure 7-1-80	XYZ Task Leader[b] and QA&I Inspector
3. Improper operation or total failure of Data Collection System Causes: (a) Malfunction of data collecting instruments, (b) Operator error	a. Develop maintenance system including schedule for recalibrating all data collecting instruments. b. Prepare special operating procedures.[a] Conduct training program for operators.[a]	Complete procedure Complete procedure 7-1-80 Complete training 8-1-80	I&C Lead Designer[b] and Y-12 Maint. Engineer

Problem/Cause	Action	Schedule	Responsibility
4. Prolonged shutdown to repair XYZ Sensors Causes: (a) Location of sensors in relatively inaccessible locations, (b) Possible high failure rate of "first-of-a-kind" sensors	a. Develop special maintenance procedure and train maintenance personnel in use of procedure.[a]	Complete procedure 7-1-80 Complete training 8-1-80	Y-12 Maintenance[b] Engineer and I&C Engineer
	b. Conduct periodic functional testing and recalibration of sensors.[a]	Complete test plans and recalibration schedule by 7-1-80	I&C Lead Designer[b] and Y-12 Maint. Engineer
5. Loss of operating capability Cause: Operator error	Develop operating procedure[a] for start-up, routine, shutdown and emergency operation of test rig. Conduct training program for all operators.[a]	Complete procedure 7-1-80 Complete training 8-1-80	XYZ Task Leader

[a]In order to assure that procedures, plans, and the training program are effective, they will be evaluated by the task leader at the beginning of operation and every 12 months (minimum) during operation.

[b]Prime responsibility.

Source: Reference 2.

REFERENCES

1. G. W. Roberts, Quality Assurance in R&D, <u>Mechanical Engineering</u>, Vol. 100, No. 9, p. 41 (1978).
2. F. H. Neill, Quality Assurance Programs in Research and Development. Quality Assurance in a Large Research Laboratory, ASQC Seventh Annual National Energy Div. Trans. Houston (1980).

Chapter 5

PROJECT DESIGN CONTROL

I. OPTIONS

Design control for nuclear construction is specified in Appendix B to 10 CFR 50 (1), and is intended to apply to all aspects of the design having an effect on the quality of safety related items. A specific emphasis is placed on reactor physics, stress, thermal, and hydraulic analyses, accident analyses, compatibility of materials, accessibility for in-service inspection, maintenance, and repair, and delineation of acceptance criteria for inspection and tests. This is certainly a major expansion of the traditional concept of quality assurance. The usual practice has been to check drawings to ensure that inspection and acceptance criteria are included, welding or other special processes are detailed and to determine whether hardware manufactured to the drawing to be inspected. A notable exception has been for defense and aerospace procurements, where formal reliability and maintainability program requirements are often imposed that require design reviews to verify the attainment of reliability goals, and accessibility to system components for in-service inspections, maintenance and repair. There is a great deal of emphasis put on actually challenging the nuclear equipment design's suitability to provide acceptable hardware to meet its intended use. Actual prototypes or first production articles may be subjected to vigorous qualification testing, or the design challenging process may take the form of formal design reviews or less formal independent technical reviews (peer reviews).

II. DESIGN REVIEWS

Historically, the initial purpose of the design review was the early detection and remedy of design deficiencies which could jeopardize the successful performance of a product or process being issued, could jeopardize the cost to build or fabricate the product, or could jeopardize the ease of maintaining that product (2).

Design reviews were initially established on a cost benefit basis. With the advent of the military and aerospace programs, however, design reviews have become more of a mandatory process and are dictated either by customer or government requirements. Specifications with design review requirements [for example, Army AMCP 703-2 (3) and NASA's NHB 5300.4(1A) (4)] are normally associated with the reliability programs. Design review is permitted by 10 CFR 50, Appendix B (1) as one alternate form of checking the design; other choices include independently verifying calculations or prototype/qualification testing.

In the past, courts have been reluctant to impose liability in design, particularly for products made by established manufacturers. They did not want a jury to have to evaluate a product prepared by experts in their field. The courts also realized that once a judgment was made against a contractor or manufacturer, it could open the door for additional claims against that manufacturer by other people. However, liability is being imposed today in an increasing number of cases for negligence of design (5). Design negligence is predicated on one of three theories (6):

1. Failure to use material of adequate strength or quality
2. The incorporation of a concealed hazard
3. The failure to employ needed or reasonably required safety devices.

In most cases today, the manufacturer is being held to the degree of knowledge of an expert. This imposes on the manufacturer the duty of an expert to know about the latest developments, including safety precautions, in the field. It is interesting to note that: "Noncompliance with a standard established by statutes, ordinance, or administrative regulation or order furnishes evidence of deficient design, and may, under certain circumstances, constitute negligence per se" (6).

The reverse is not necessarily true. Mere compliance with codes and standards may not prove the manufacturer has exercised sufficient reasonable care in designing a safe product. Therefore, compliance with codes should not be taken as a substitute for a design review.

Although the initial objectives of a design review program may be to minimize the risk and cost associated with new products or processes developed by the R&D division, the use of the design reviews could be extended to new or redesigned experimental facilities within the division. Identification of the need for this type of design review would be on a case-by-case basis, as determined by the appropriate laboratory manager. While the provisions for formal design reviews are offered as guidelines, the actual reviews for experimental facilities could be conducted at varying levels appropriate for the particular situation.

The formal design review process generally includes subjecting the design to at least two reviews. The first review is usually during the initial stages of planning to concur with the design specifications. The second review is performed at the completion of the design and prior to construction of the facility or transfer of the product or process for manufacturing. Each of these reviews are conducted by technically experienced and skilled personnel from within or outside of the division, and who are not directly associated with the development of the design being reviewed. At the conclusion of both specification and design reviews, the recommendations of the review board should be documented and signed by all participants, noting an understanding of the recommendations, and including the opinions of participants who disagree with the recommendations. Additional specification review meetings should be scheduled as necessary if modifications to the specifications are required. These review meetings, particularly the meeting which has as its purpose the final approval of the design, rely on a number of characteristics to ensure their success. The first of these is the element of formality. Design reviews should be planned and scheduled like any other legitimate activity. The meeting should be structured around some sort of agenda, which typically could include the following:

1. Review of the specifications for the product, process or service under consideration
2. Stress and structural requirements
3. Codes and standards
4. Environmental requirements
5. Maintenance consideration
6. Materials of construction
7. Construction features

Design review checklists will vary depending on the industry or product or service involved, but the foregoing is illustrative of the scope of the review. A copy of this agenda should be sent out to the design review team members prior to the meeting, along with sufficient additional information for them to become familiar with the design characteristics and to be able to develop pertinent questions to be discussed during the design review meeting. The design information package should at least include the agenda (as described above); alternatives to the design; results of any special tests performed to substantiate the design or in conjunction with the design; the design drawings, the recommendations developed during the specification review; a comparison of the design to the specifications; costs to manufacture to the recomended design; and costs for alternative designs which have been considered. Finally, include minutes from any previous review meetings, if applicable.

During the meeting of the design review board, minutes should be generated and subsequently approved by all members of the design review board. After completion of the board's review, a final report should be prepared and distributed to the board members and any interested parties. The design review report should include a discussion of the objectives of the design, specifications, design drawings, backup information, and recommendations of the design review board, including minority opinions. Once the design review report has been generated, it should be signed by all members of the design review board signifying their agreement with the contents of the report. A file should be maintained for all information pertinent to the design review process, including the information package that has been distributed to participants and the documentation of the review meeting. The design review report and backup file should then be retained for as long as the item is being marketed or used.

III. INDEPENDENT TECHNICAL REVIEW

In commercial applications, if all designs were required to be verified or challenged, the use of formal design reviews would be very costly. For the R&D center, a more workable solution is a system whereby the reviewing of technical data is done on a day-to-day basis involving one or two people, rather than an entire review team with formal meeting agendas, minutes, etc. The application of a formal design review is contingent upon the number of areas impacted by that design or design change, and is a function of the relative importance of the design. For the launching of a major new product, several organizations are represented in a formal meeting with senior management. In a research project involving qualification testing of one or two components, one person not having worked on the data but capable of doing an independent review can accomplish the task more easily and efficiently. A research center performs in-depth design analyses for a field of interest which may be less broad than the total design of a given product; for example, vibration analysis of a tube bundle assembly within a large steam generator. The independent technical review is oriented toward that tube bundle analysis, and not toward the overall steam generator, at least as far as the research center is concerned. The results of that analysis and several additional analyses over different components of that steam generator are then taken into account during a major design review activity for the overall steam generator at a later time. Taking this reduced scope into consideration, the requirements for independent technical review for a component can be modified from the requirements for design review at the system level.

The general requirements for a nuclear design control program include identification of design interfaces, coordination among the participating design organizations, and translation of the design basis into a design document (1). The project technical plan (see Chapter 4) identifies the technical requirements and objectives of the research project. The project technical plan includes the philosophy and basic principles involved in the choice of the test or experimental design concept, and any inherent limitations of that design concept. Equipment designs or calculations which provide a basis upon which the experiment is to be conducted, the data is to be evaluated, or the final conclusions are to be drawn should be documented and subjected to an independent review by a competent individual, other than the one who performed the original design or calculation. If a computer program is used, the input, output, and applicability of that program is verified and the verification documented (see also Chapter 10). Using the originator's supervisor to perform an independent review should be restricted to special situations where the supervisor was the only individual within the organization competent to perform this review. The overall scope of an independent review should include the project leader's calculations, which were made based upon the customer's inputs (customer means the organization sponsoring the research, whether from inside the corporate structure or outside as contracted research), the design engineer's calculations and drawings (the term design engineer is defined as the person who designs the test fixture or test articles, as opposed to the project leader who is responsible for the overall concept of the experiment), and the project leader's calculations required to complete the final report.

If the design of the test section is done by a separate organization other than the project leader, the design engineering calculations and drawings should be submitted to the project leader for an independent review. The project leader may elect to do the review himself or select a reviewer to do that verification function for him. The project test section design should be evaluated for:

1. Conformance with project specifications;
2. Material compatibility with the environment, such as pressure, temperature, corrosion fluids, water chemistry, etc.;
3. Design interfaces, as related to the fitup of parts;
4. Dimensional stability, including the coefficient of expansion, creep, fatigue, loading, etc.;
5. Conformance with plant safety requirements;
6. The correctness of information used for the calsulations;
7. Adequacy and correctness of any computer programs involved; and
8. Appropriateness of specified quality standards.

There are certain documentation suggestions which will make it much easier for an auditor to ascertain that an appropriate independent review was performed. The QA organization can then verify that the review system is in effect without necessarily getting into the technical review process themselves. In any case, of course, if there is a question as to the adequacy of the independent review, the auditor must have the option to bring in a competent third party to perform another independent review, therefore verifying the adequacy of the original review.

A. Documenting Calculations

Calculations are documented as backup evidence for completion of research (see Figure 5.1). The calculations should be identified by the project number and title, and should include date performed, originator's name, and a cross reference to the drawing specification or the system or test phase to

PROJECT TITLE P-1 FLOW MODEL TEST   **CALCULATION**
PROJECT NO. 9999   **SHEET**
SUBJECT CONVERSION OF THERMOCOUPLE OUTPUT TO TEMP. PAGE 1 OF 1

The equation form used to relate the thermocouple output voltage to temperature is:

$$T = C_1 V^N + C_2 V^{N+1} + \ldots + C_6 V^{N+5}$$

WHERE : T = temperature (°F)
V = thermocouple output voltage (mv)
C_{1-6} = calibration constants
$N = 0$

The following calibration constants were obtained by curve fitting the conversion tables for type T thermocouples using the polynomial equation given above.

The conversion tables used were taken from NBS Circular 561 by the L&N Co. The curve fit was obtained using the program POLYAA.

$C_1 = 32.086$ $C_4 \ -0.64147$
$C_2 = 46.348$ $C_5 \ +0.26500$
$C_3 = -0.55101$ $C_6 \ -0.0356140$

The results are valid only over the range of

$$0 \le T \le 150°F$$

Results valid for 32°F reference

PERFORMED BY BBLukas DATE 12-1-81 CHECKED BY N.A. DATE
(ASME CODE APPLICATIONS ONLY)
REF. DRAWING NO. N.A.   REF. LOG BOOK PAGE N.A.

Fig. 5.1. Calculation sheet. (Courtesy of Babcock & Wilcox.)

which the calculations apply. The calculations should be entered in a manner that is legible and in a form suitable for reproduction. They should be sufficiently detailed as to the purpose, method, assumption, design input, references and units, such that the person technically qualified in the subject can review and understand the calculation and verify the adequacy of the results without contacting the originator (7). This detail should typically include:

1. Identification of the objective of the calculation;
2. Identification of the design inputs and their sources;
3. Documentation of assumptions and identification of those assumptions that must be verified as the calculation proceeds;
4. Functional performance requirements: What is the range of the parameters to be investigated?
5. Quantitative values necessary for defining fabrication, installation, or testing;
6. Results of the analysis: Do the results conform to the objectives established in the project technical plan?
7. Conclusions: How do the results compare to the predicted values supplied by the operating division?

B. Documenting Independent Reviews

In a like manner, when performing the review of calculations or designs, certain disciplines should be followed for the documentation of those reviews (see Figure 5.2). The reviewer should indicate the project number and title, the identification of the person performing the review, and the specific data being reviewed, whether it is laboratory notebook pages, calculation sheets, or computer programs, etc. The reviewer should indicate the approach used for performing the review and the results of the review; i.e., the acceptability of the data being reviewed. Questions concerning the data should be answered by the originator to the satisfaction of the reviewer. The reviewer may elect to verify that the method used for calculations used is correct, and by any method suitable to him, may recalculate all or a sufficient portion of the data to verify its correctness.

IV. DRAWINGS

Thus far in the discussion on design control, we have been talking generally about reviewing the design with respect to the overall project goals, the adequacy of the experimental design, the adequacy of the calculations used for proper scaling of the experimental models, or those calculations used to reduce experimental data. Certainly in any discussion of design review, the matter of drawings used for test sections should be addressed. Test

RC-366

PROJECT TITLE _HEAT TRANSFER FACILITY_

PROJECT NO. _2067_

SUBJECT _SINGLE PHASE VENT HEAT EXCHANGER_   PAGE _3_ OF _3_

REVIEW

5. CHECK h_{TUBE}

$$h = \frac{k}{D} = 0.0155 \, Pr^{.5} \, Re^{.83} \qquad (KAYS \, p.173)$$

ASSUME $\overline{Pr} \cong 1$

$$h \cong \left(\frac{.402}{.0417}\right)(1)(150,000)^{.83} \, .0155$$

$$h \cong 2955 \; {}^{BTU}/_{HR-FT^2 \, °F}$$

6. CHECK U

$$U_{OD} = \frac{1}{\frac{.5}{.37}\left(\frac{1}{2955}\right) + \frac{.25}{(12)(9.76)} \, \ell_n\left(\frac{.5}{.37}\right) + \frac{1}{1931}} = \frac{1}{.0005 + .0006 + .0005}$$

$$U_{OD} = \frac{1}{.0016} = 618. \; {}^{BTU}/_{HR-FT^2 - °F}$$

7. CHECK A

$$q = U A \Delta T = U A L_{MDT}$$

$$A = \frac{q}{U \, L_{MDT}} = \frac{717,090 \; {}^{BTU}/_{HR}}{(618 \; {}^{BTU}/_{HR-FT^2 \, °F})(176 °F)} = 6.59 \, FT^2$$

$$A = \pi d L \Rightarrow L = \frac{6.59 \, FT^2}{\pi \, .0417} = 50.3 \, F.T.$$

8. CONCLUSION

A HEAT EXCHANGER OF 50 FT. PLUS 25% OR

~63 FT. SHOULD BE ADEQUATE FOR THIS

APPLICATION.

PERFORMED BY _Paul J. Rozich_   DATE _8-10-83_

REFERENCE SOURCE _KAYS p.173_

Fig. 5.2. Review sheet for calculations. (Courtesy of Babcock & Wilcox.)

section designs are highly fluid, particularly during the early phase of the project, and there must be provisions built-in for rapid changes to the drawings used for test section construction.

The technical group responsible for the research may include engineers who are quite capable of generating their own drawings. On the other hand, the group may be of a scientific discipline with no expertise in drafting. Therefore, each group may have its own ideas about how to (or whether to) generate its own drawings in the interest of keeping the project moving rapidly and cheaply.

There is, however, a very pressing need to establish a standardized technique for generating and revising drawings. This will ensure that any revisions to these drawings are approved and transmitted to all who need them. Some sort of centralized drawing control is needed, regardless of whether one organization does all the design or whether the project leader does some drafting himself and some is done by others at the research center. Once the drawing has been completed, it should be routed to a central-

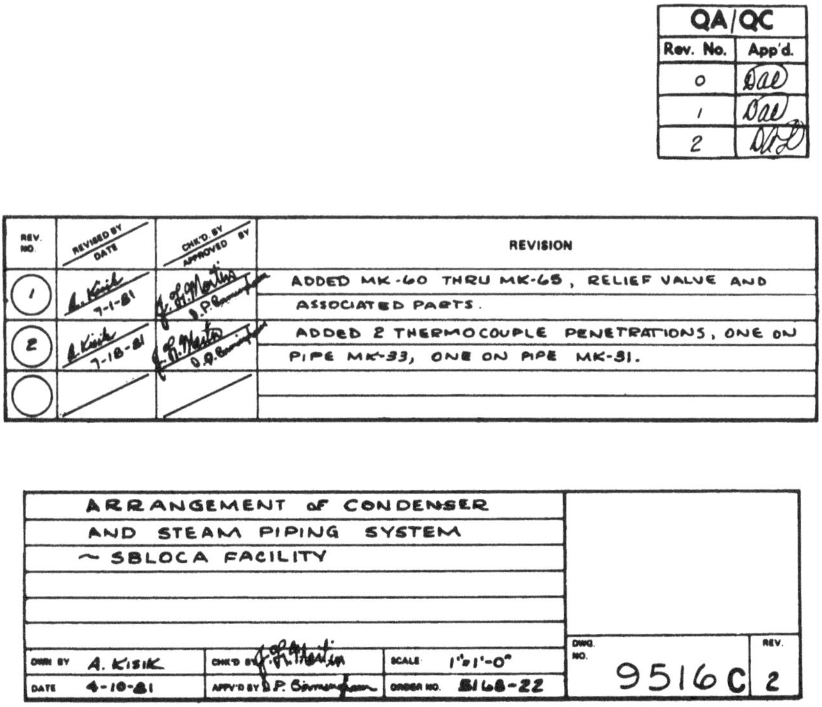

Fig. 5.3. Drawing sign-off block. (Courtesy of Babcock & Wilcox.)

ized location to ensure that appropriate drawing numbers are assigned and
that the drawing reviews have been performed. These drawing reviews (see
Figure 5.3) should include a review by someone in the design organization
knowledgeable of state and local building codes, and a review by QA profes-
sionals to ensure inspectability and to help with dimensioning. All drawings
should be approved by the project leader.

REFERENCES

1. United States Code of Federal Regulations, 10 CFR 50, Appendix B,
   Quality Assurance Criteria for Nuclear Power Plants and Fuel Reprocessing Plants, Government Printing Office, Washington, D.C. (1970).
2. J. M. Juran, Quality Control Handbook, McGraw-Hill, New York (1974).
3. AMC Pamphlet No. 702-3, Quality Assurance Reliability Handbook,
   Headquarters, U.S. Army Materiel Command, Washington, D.C.
   (1968).
4. National Aeronautics and Space Administration NHB 5300.4(1a), Reliability Program Provisions for Aeronautical and Space System Contractors, (Formerly NPC 250-1), U.S. Government Printing Office, Washington, D.C. (1971).
5. Product Liability, The Present Attack, American Management Association (1970).
6. Corpus Juris Segundum 72 CJS Supplement 21, Products Liability, West
   Publishing Co., St. Paul (1975).
7. American National Standards Institute ANSI/ASME NQA-1-1979, Quality
   Assurance Program Requirements for Nuclear Power Plants, The American Society of Mechanical Engineers, New York (1979).

Chapter 6

PROCUREMENT CONTROL

I. PROCUREMENT DOCUMENTS

A. Identify Critical from Noncritical

A key element of the procurement cycle is the procurement document. It may be called a purchase requisition or a purchase order, or in some cases, a combination of forms all having the purpose of specifying the product or service to be purchased by the research center, along with appropriate technical and quality requirements to be met by the supplier. Many of the materials to be purchased for a project for R&D may not be actually critical to the quality or validity of the project data, but there are a sufficient number that are critical to warrant a formal system for review of procurement documents. This review should be performed by QA since they are familiar with the overall aspects of supplier evaluation and control and knowledgeable of the consequences of not adequately specifying material type, composition, markings, applicable codes, etc. At the time the project technical plan and the QA plan are developed, the project leader and the QA representative should have a general idea of which procurements will be critical to the results. If possible, they should be identified for later reference when the purchase requisitions actually are generated. Purchase requisitions should be identified to the job identification number to allow a reviewer to trace back to the basic technical plan and QA plan requirements.

The initial decision as to whether a procurement is subject to QA can only be made by the requisitioners unless there are provisions for QA to review all procurement documents. This is highly unlikely in a small research center, so the project leader must be charged with the responsibility of indicating on the procurement requisition or purchase order that QA is involved, possibly by checking a "yes" or "no" block (see Figure 6.1). Purchasing should make sure that the block has been checked either "yes" or "no." If the block is checked "yes," then the procurement should have been reviewed by QA.

**PURCHASE
REQUISITION NO.**

DATE WRITTEN	DATE REQUIRED	APPROX. COST ESTIMATE	CHARGE NO.
8-28-81	9-15-81	$50.00	2011-01

REQUESTED BY	DELIVER TO	
F. White	Instruments	☒ YES, ☐ NO, QUAL. ASSUR. REQ'D. ☐ YES, RADIOACTIVE MATERIAL ☐ YES, RADIATION PROD. EQUIP.

CAN BE PURCHASED FROM THESE VENDORS:

1. XYZ Labs — DOES 10 CFR 21 APPLY? ☐ YES ☒ NO — P.O. NUMBER

2. Acme Labs — CODE MATERIALS ☐ YES ☒ NO

V E N D O R — EXPEDITE ☐ / EXPEDITE DATE / P.O. DATE

VENDOR PHONE NO. / VENDOR CODE / B.C.

CONFIRMING ORDER ☐ YES TO: / TAXABLE ☐ YES ☐ NO

✱ THE DATE LISTED BELOW IS THE DATE ITEMS ORDERED ARE REQUIRED TO SHIP TO DESTINATION — SHIP VIA

F.O.B. ☐ S/P ☐ DELIVERED / TERMS

✓ITEM	QUAN.	UNIT	NOMENCLATURE	*SHIP DATE	COMCOD#	UNIT PRICE	EXTENSION
1	1	ea.	Boeckeler Model 10 CR Spindle Micrometer				
			Head, B&W Serial No. 630111, to be re-				
			calibrated.				
			Note:				
			(1) Calibration required to nearest				
			.000010 inch.				
			(2) Certification required as set forth				
			on page 2 of this requisition.				
			APPROVED SUPPLIER REQUIRED (QUALITY EVALUATED)				
			X – P.O.D. YES☐ NO☐				

DOCUMENTATION: FILL IN BELOW WHEN REQUISITION EXCEEDS $1,000.00, AND:
1. ONLY ONE SUPPLIER'S PRODUCT OR SERVICE IS APPROVED.
2. TO BE PURCHASED FROM OTHER THAN LOWEST BIDDER.

TOTAL ►

REQUISITIONER

REASON:

DEPARTMENTAL APPROVAL

COMMENTS: CC: QA, Instruments, Applied Mechanics

BUYER, PURCHASING DEPT.

ORIGINAL

Fig. 6.1. Purchase requisition. (Courtesy of Babcock & Wilcox.)

B. QA Review

The QA review would include verifying the inclusion of basic technical requirements, provisions for access to the supplier's facility for a source inspection or audit, documentation submittal and retention requirements, general requirements for establishment of a QA program at the supplier's facility, provisions for extending applicable requirements to lower tier suppliers, and requirements for material certifications or certificates of conformance. Quality Assurance would also verify whether the supplier had been formally evaluated and approved, either by the R&D center, another company division, or a third party such as the Coordinating Agency for Supplier Evaluation (CASE (1) or American Society of Mechanical Engineers (ASME) (2). These requirements are similar to those imposed on production type operations. The difference from an R&D standpoint is the criteria depends on whether the procurement affects the quality or validity of the data.

Of procurements that can have an effect on research, the procurement of data-taking devices is of significant importance. Therefore, evaluation and control of suppliers of instrumentation becomes a significant issue. Since the normal output of the research center is data, the experimentation, data collection, and analysis functions of the research center are areas of critical importance. The procurement documents for instrumentation and gages for the research center should be reviewed.

C. Specification of Requirements

The supplier should be told specifically what is wanted, ideally before he is asked to ship the order. Although this sounds self-evident at first, in many cases purchasers are in such a great rush to receive their material, they will place the order over the phone and state that the purchase order will follow later. The supplier assumes he has all the information, proceeds to ship the order (usually taking the item from stock), but finds out when he receives the actual purchase order that some specific requirements have been added. Then the time saved by the verbal purchase order and quick shipment is lost because the material is held up for resolution of the discrepancy. Oftentimes, purchase orders are quite vague about what they're asking the supplier to provide. Casual statements such as "certifications required" do not tell the supplier very much. What he needs to know is what kind of certification — does it involve a chemical analysis or physical test, or a certificate of conformance to a particular specification?

Purchase orders should be carefully reviewed to ensure they specify the material or item by appropriate ASME or ASTM (3) designation or catalog number, material grade or size, performance rating, pressure rating, or other functional or physical characteristics that are necessary. Quantity, length and weight, and identification of the test reports required should be spelled out carefully. When citing the procurement to a specific code, be sure to include the revisions or addenda of that code. The same rigor applies when purchasing calibrated equipment or calibration services. Just because the research center may be operating to some calibration specification, such as MIL-STD-45662 (4), do not assume that suppliers who calibrate measurement standards are also operating to the same specification. Nor should it be assumed that a certification of traceability to the National Bureau of Standards is any more valid for a calibration service than a certificate of conformance is for materials. For some levels of activity, these paper certifications are sufficient to establish some degree of confidence. But, for highly critical experiments with high cost or safety related consequences, a more thorough investigation of the supplier's capability behind the certificate is warranted. For that reason, requirements for approved suppliers and specific technical and administrative requirements to assure quality should be included in procurements for calibrated equipment and calibration services.

1. Calibration Sources Although no national consensus standard as yet exists which may be appropriate to the non-military calibration laboratory, specific requirements should be levied to assure the quality of their work since it so directly affects the quality of the data at the research center. A basic checklist of those requirements might include a requirement for the supplier to list the calibration equipment used at the supplier's facility by serial number and a statement to the effect that those items were calibrated traceable to the National Bureau of Standards. It would then be possible at a later date for the research center to verify by audit that those supplier standards were maintained traceable to the National Bureau of Standards. As discussed in Chapter 8, "traceability" for a research center requires traceability of the data, not just an equipment check against a standard claimed to be 4 or 10 times more accurate. Actual numbers must be on file with the supplier to substantiate a numerical uncertainty statement. In those instances where units are being sent back to a supplier for recalibration, a check should be made by the supplier "as received" before any repairs of adjustments are made. A precalibration report furnished to the research center with the data points measured enables the research center to determine whether the equipment drifted out-of-calibration during use. The effect of that condition on any previous data that was taken by the instrument can then be assessed. If no repairs or adjustments are made to the unit, this precalibration check can then be used as the final calibration report. If there are repairs or adjustments made to the unit, the supplier then needs to furnish a final calibration report showing the final readings taken with the instrument. Finally, the supplier's procedures for the calibration should be referenced on the calibration report. The research center then has the option of reviewing those procedures later if they did not do so before the supplier was placed on the Approved Suppliers' List. Although many commercial calibration labs use standard calibration manuals supplied by their corporate headquarters, there may be a tendency on

the part of the technicians in the local areas to modify the calibration steps to suit themselves. These modifications are not always condoned or authorized by the supplier's management. For that reason, it is prudent to require the supplier to use a written procedure which has evidence of management approval and to reference that procedure on the calibration report.

The formal specification of these requirements on a purchase order may come as a shock to many suppliers who have not been in business in the military, aerospace or nuclear markets. It is an important function of a survey team to explain these requirements to prospective suppliers so they can understand the impact of the requirements on their operation.

2. Suppliers of Research For many of the same reasons given in Chapter 1, any organization anticipating the procurement of research should be concerned with the quality of the service. Like any process, the quality to be designed and built into the deliverable end item is not measurable at the time of delivery. The basis for achieving quality must have been laid from the beginning and adhered to throughout the entire project. Quality is not applied retroactively by auditing the paper at the end of the project.

It is extremely important that there be a meeting of the minds between customer and supplier as to what is going to be accomplished. Too often, the customer project leader and the supplier project leader will become so completely engrossed in following the project technical details that they forget to pay attention to business. The money runs out before the test is complete and both sides wonder what happened. Then the finger pointing begins.

A statement of what is to be done and the verification and documentation to be incurred should be developed at the very beginning and adhered to throughout the project by both parties unless common written agreement is reached spelling out what changes are needed and what the cost and schedule impact will be. This is usually contained in the project technical plan and QA plan and can be as brief or as lengthy as necessary to adequately cover the task.

Appendix A includes a suggested specification that might be useful when dealing with research centers. Its provisions apply only to those areas affecting quality or validity of data — a judgment that is made case-by-case by both parties. Rather than the customer taking a great deal of time and effort to spell out the specific details of the research project, it gives the supplier latitude to interpret what he perceives the customer's needs to be in terms of the supplier's capabilities and willingness to perform. Since the supplier is generating the document, the supplier can provide a detailed description of the facility to be constructed or used and renders his opinion as to what is critical to the experimental results. The project technical plan is, therefore, tailor-made to the supplier's terminology and technology. The customer still specifies the basic test parameters and exercises approval over the project details contained in the technical plan, but stays away from unnecessary involvement in noncritical facility rearrangements except as they might affect project costs.

Depending on the amount of control to be exercised by the customer and whether a supplier has a QA function, the requirements and commitments for verification and documentation of test articles and results can be included in a separate QA plan or written in with the project technical plan.

The decision to impose independent technical review of the test results and calculations depends on the reason for subcontracting the research in the first place. If it was because the customer just didn't have the time or the facility available but has the technical expertise, a sufficient review may be performed by the customer provided there is access to the raw data taken and the customer can witness the data collection process. As with all quality related activities, the more remote the customer is from the "action", the more reliance must be placed on a supplier's internal independent QA function.

II. SUPPLIER EVALUATION

The R&D center may have access to a corporate Approved Supplier's List or various lists developed by other operating divisions if the research center is a part of a large corporation. Using prior developed lists saves a lot of money in evaluating and developing unknown suppliers to meet your particular needs. Purchases placed through a central purchasing allows for combining needs of several facilities and the advantage of bulk rates and more leverage on a supplier to meet your special needs. The disadvantage is that R&D centers may only need a small piece here and there, and cannot wait the long lead times required for large quantity orders. The smaller orders are placed with local retailers who can turn the order out within 24 to 48 hours. However all too often, the local retailer will not have material certifications or will charge an exorbitant price to provide them. Then, what may be received is a certification typed on the retailer's letterhead instead of the letterhead of the company who actually performed the test. For many codes, this is "verboten." A goal for the vendor evaluation and selection process for a research center is to find those suppliers who can provide special order items on fairly short notice, and who are willing to provide the proper certifications, all for a reasonable price — no small accomplishment indeed. When developing new suppliers, the research center should give consideration to establishing centralized supplier history files. The purchasing organization often has such files, but just as often, they are limited in their scope with regard to the prices and achievement of delivery schedules. This is excellent information in evaluating a supplier's performance, but it doesn't go far enough. Quality problems should also be included in the centralized files and they should be used as a basis for evaluation and selection of a supplier for an Approved Suppliers' List.

Project leaders should be encouraged to put their experiences with their suppliers into the central history files. Often during the course of a research project, a project leader will have several interactions with a supplier of specialized test equipment. The results of those interactions, both positive and negative, should become a part of the research center's data bank. All too often, the results of these transactions are kept in the project leader's personal files and these are unknown to anyone else at the research center. Since supplier capabilities and limitations may vary from one product line to another, particularly where different physical plants are involved, centralized evaluation files based on the supplier's overall performance will allow the comparison of different experiences with the same suppliers by different project leaders, and highlight the relative strengths and weaknesses within the same supplier structure. Some suggested criteria for evaluation and selection of suppliers are:

1. The supplier should have a good record of supplying acceptable items. The research center may use supplier history data obtained from other sources, such as corporate Approved Suppliers' Lists or other division's supplier quality history. The history should be for similar items, however, so the comparison remains valid. If the purchase is of a complex or high cost item, the supplier should be evaluated by the technical section projett leader and QA to determine the capability of the supplier to perform.

2. For some procurements, the research center may wish to perform an audit or, alternatively, to accept the supplier on the basis of a certificate of accreditation; for example, American Society of Mechanical Engineers' (ASME) Quality Systems Certificate for Materials; ASME Code Stamp Certificates for Fabrication of Boilers and Pressure Vessels; National Voluntary Laboratory Accreditations; or an evaluation performed by consolidated evaluation groups, such as the Coordinating Agency for Supplier Evaluation (CASE).

3. For commercial or off the shelf items, a receiving inspection may be sufficient to determine compliance with the procurement document requirements. This is somewhat risky, however, since you have to wait until you receive the product in hand before you know whether the supplier is any good. An alternate consideration would be to perform a source inspection at the supplier's facility prior to permitting the supplier to ship the item to the research center. If this is to be done, the supplier should be notified well in advance and given the specifics as to when the research center wishes to inspect the product, whether it's during the course of the fabrication process or at the completion of the fabrication process just prior to shipment. The advantage of this, of course, is that if there is something wrong with the item, the supplier has it at his facility and taking up his storage space instead of the research center's storage space. Additionally, the materials and personnel are available immediately to start correcting the item if correction is possible.

III. RECEIVING INSPECTION

Once the purchased item is received at the research center, it should be immediately inspected to verify that the material does in fact meet the purchase order requirements. This inspection may be by a receiving inspector if such an inspection is warranted, but at the very minimum, it should be done by the project leader or one of his technicians. The receiving inspection should be documented (see Figure 6.2) and should require immediate corrective action if there are any discrepancies with the material. This is to permit the research center to collect for any adjustments due to defective material, and to prevent schedule and construction delays later on down the line if the material is pulled out of stock and found to be unserviceable. The purchase order should not be closed out nor should the supplier be paid until the receiving inspection is performed and signed off. Appropriate tags should identify the material being held for inspection, that which has been inspected and accepted, and that which is discrepant.

When setting up the system for procurement control at the research center, much effort must be expended in coordination meetings and discussions with all affected parties to assure that everyone understands the requirements of the system. This means getting together those persons that are primarily involved in the requisitioning process, the project leaders, designers, and the purchasing agents, and going over the details of the procurement control system item by item with appropriate examples, overheads, viewgraphs, or whatever is necessary to make sure everyone understands the system and its salient features. This is a prime responsibility of the QA organization, because unless this understanding is achieved from the very outset, problems will crop up during the implementation phase beyond those that can normally be expected with instituting any new system.

APPENDIX: SUGGESTED SPECIFICATION FOR QUALITY ASSURANCE
                 REQUIREMENTS FOR RESEARCH AND DEVELOPMENT

This specification defines the Quality Assurance Requirements for the performance of R&D programs. The requirements of this specification apply to the development of the design of the experiment or test through detailed design of the test article, fabrication and data acquisition, to reporting of the results and maintenance and disposition of associated records.

THE REQUIREMENTS OF THIS SPECIFICATION SHALL APPLY ONLY TO THOSE PORTIONS OF THE R&D PROGRAM OR PROJECT THAT AFFECTS THE QUALITY OR VALIDITY OF THE DATA OR RESULTS.

RECEIVING INSPECTION REPORT

CHARGE NO.			PAGE	OF
6742-02			1	6

DRAWING NO.	REVISION NO.	PURCHASE ORDER NO.	
1050-E	1	10034-04	

PREPARED BY:		DATE	APPROVED BY:		DATE
Frost		3/30/81	*a. R. Boss*		4-7-81

INSPECT	LINE ITEM	QUANTITY	CATALOG NO.
	1	10 Sections	1075-160

DESCRIPTION

3/4" Sch. 160 Pipe

ACTUAL OPERATING CONDITIONS	SPECIAL CONDITIONS
2900 PSIG AT 650°F	

CHARACTERISTICS (TO BE FILLED OUT BY REQUISITIONER)			INSPECTOR RECORD DATA — DATA MUST BE ENTERED FOR ALL CHARACTERISTICS ENTERED BY REQUISITIONER	A C C	R E J	INSP BY	DATE
MATERIAL SPECIFICATION SA/SB			SA/SB 312 TP 304	☑	☐	JaB	4-17-81
			HT # 465478	☑	☐		
ASME SA 312, TP 304			MFG SANDVIK	☑	☐		
PIPE	1 OUTSIDE DIA.	1.05	1.054	☑	☐		
	2 WALL TK	Av. 219-Min. 192	.219 to .222	☑	☐		
	3 SCHEDULE	160	160 seamless	☑	☐		
	4 LENGTH	210 ft. ± 20 ft.	10 sections 223 FT	☑	☐	JaB	4-17-81
PLATE	5 THICK			☐	☐		
	6 SIZE OR DIA.			☐	☐		
OTHER	7 DESIGN CONDITIONS			☐	☐		
	8 MFG. DESIGNATIONS			☐	☐		
	9 SIZE DESIGNATION			☐	☐		
	10 MATERIAL			☐	☐		
MILL TEST REQUIRED YES ☒ NO ☐			Mill Test Report on File	☑	☐	JaB	4-21-81
CERTIFICATE OF CONFORMANCE FOR: MATERIAL ☐ PRESSURE ☐ TEMP ☐				☐	☐		
OTHER MARKING, TAGS, INFORMATION			ENTER ANY ADDITIONAL BY REFERENCE NO.				

MEASURING DEVICE USED	SERIAL NUMBER				
VERNIER CALIPERS	790024				

ACCEPTED	DATE 4-21-81	QUANTITY 223 FT	**REJECTED**	DATE	QUANTITY	**DEVIATION REPORT**	DATE

REMARKS

ACCEPTED BY	*J. A. Brown*	DATE 4-21-81

Fig. 6.2. Receiving inspection report. (Courtesy of Babcock & Wilcox.)

Definitions

<u>Certificate of Conformance</u>	A written statement signed by a qualified party certifying that items or services comply with specific requirements.
<u>Certified Test Report</u>	A written document approved by a qualified party containing sufficient data and information to verify the actual properties of items and the actual results of the required tests.
<u>Project Technical Plan</u>	A written document prepared by the Research Organization that includes the type of experiment, parameters to be investigated, work scheduled, order value, a general description of how the test section or experiment is to be constructed, identification of materials having an effect on the test results, the general measurement program to be used and the types of data to be collected.
<u>Research Organization</u>	Any organization under contract to perform a specific research and/or development project.
<u>Purchaser</u>	The organization that contracts for a research or development service from a Research Organization and is the organization invoking this specification as a condition of that contract.

<u>Section</u>   <u>Organization</u>

1.0   <u>ORGANIZATION</u>
A quality assurance organization with the authority and responsibility for establishing, planning, and implementing the required quality program is required. No single organization pattern is mandatory. However, an organized approach demonstrated by appropriate organizational charts and written descriptions which clearly define the authority and duties of all persons involved in the quality program is required. Quality assurance personnel shall have sufficient organizational freedom to initiate, require, and verify resolution to quality problems. This freedom shall include sufficient authority to control

further processing and testing of a nonconforming item, deficiency or unsatisfactory condition until proper dispositioning has occurred. In addition, the person or organization responsible for the quality program shall have a direct access to responsible management at a level where action can be taken, shall be sufficiently independent from the pressures of production, and shall report at least monthly to responsible management on the effectiveness of the program.

2.0   QUALITY ASSURANCE PROGRAM
A QA plan that describes the method for implementing the requirements of this specification shall be submitted by the Research Organization to the purchaser together with a Project Technical Plan. If it is not feasible to submit the entire QA Plan with the Technical Plan, a schedule for submittals of the remaining items shall be provided. The portions of the plan applicable to the area of work to be performed must be submitted and approved prior to the commencement of the affected work. The details of the program implementation shall be included in the QA Plan or referenced to a specific section and revision of the Research Organization's QA Manual. Document/data submittal requirements shall be evaluated individually for each project and shall be identified in the QA Plan.

3.0   DESIGN CONTROL
3.1   Justification of Design Concept
The philosophy and basic principles involved in the choice of the test or experiment design concept will be included in the Project Technical Plan. The inherent limitations of the design concept will be identified by the research organization.

3.2   Independent Review
The Research Organization shall assure that equipment design or calculations which provide a basis upon which the experiment is to be conducted, the data is to be evaluated, or the final conclusions are to be drawn, are documented and subjected to an independent review by a competent individual other than the one who performed the original design or calculation. If a computer program is used, the input, output and applicability of the program will be verified and the verification documented.
In the case of design documents, the use of the originator's supervisor to perform this review should be restricted to special situations where the supervisor is the only individual within the organization competent to perform this review. Justification for such use should also be documented along with the extent of the supervisor's input into the design document.

3.3   Data Reduction and Analysis
The method for reduction and analysis of data shall be included in the Project Technical Plan. Calibrations and all computer codes used and any discussions on doubtful scaling parameters and their resolution will be included in the final report.

4.0 PROCUREMENT DOCUMENT CONTROL

4.1 General

The Research Organization shall be responsible for assuring the adequacy and quality of all items or services supplied by their sub-tier
vendors. The procurement documents for items or services shall include the applicable requirements of this specification and of the purchaser's Procurement Documents. Each procurement document shall
be reviewed by a member of the Research Organization QA organization to assure that all applicable requirements are included.

5.0 INSTRUCTIONS, PROCEDURES, AND DRAWINGS

5.1 General

Activities affecting quality or validity of the data or results shall be
prescribed by written instructions, procedures, or drawings as required by the applicable paragraphs of this specification and shall be
accomplished in accordance with these documents.

These documents shall be submitted to the purchaser as agreed upon
between the purchaser and the Research Organization and shall be
identified in the QA Plan.

5.2 Documents

Documents affecting the quality or validity of the data or results, such
as test equipment drawings, including assembly fabrication and model
drawings, inspection checklists, technical procedures, and final reports, shall be reviewed for accuracy and conformance to test or experiment requirements.

6.0 DOCUMENT CONTROL

6.1 Changes

Any changes to the project technical plan, procedures, drawings, inspection checklists, QA plans, etc., shall be documented as to the
changes and reasons for the changes. The revised documents shall be
reviewed in the same manner as the original prior to implementation.

6.2 Control

The Research Organization shall have a system to assure that only
proper revisions of documents are utilized in design fabrication and
testing.

CONTROL OF PURCHASED MATERIAL, EQUIPMENT, AND SERVICES

7.1 General

The Research Organization shall assure that purchased items and services, whether purchased directly or through vendors, conform to the
requirements of this specification and the purchaser's sub-tier procurement documents, as applicable.

7.2 Evaluation and Selection

Each selected vendor shall satisfy one or more of the following
conditions:

1. The Vendor shall have a previous and continuous record of supply-
    ing acceptable items, processes or services of the type and quality

required by this specification and the purchaser's procurement documents.

2. An audit of the Vendor's facilities and quality assurance system shall indicate that he is capable of supplying items, processes, or services which will meet the requirements of the Procurement Documents.

3. When commercial or off-the-shelf items are procured from a "jobber" or other sources where no quality control inspection exists, the Research Organization shall perform a receiving inspection to determine conformance to the requirements of the Procurement Documents.

7.3 Source Inspection and Receiving Inspection
Receipt Inspection shall be performed to verify the requirements of the procurement documents have been met. This shall include physical inspection of the item and review of the documentation such as test reports. Source inspection shall be performed as deemed necessary by the Research Organization and as documented in the Quality Assurance Plan.

8.0 IDENTIFICATION AND CONTROL OF MATERIALS, PARTS, AND COMPONENTS

8.1 General
A list of selected materials, parts, and components that could affect the validity of the test results shall be included in the project technical plan. This list shall indicate if certified test reports or certificates of conformance are required. Any additional testing on the material that will be performed at the Research Organization shall be noted.

8.2 Identification
The identification of items affecting the quality or validity of the data or results, shall be maintained through storage, fabrication, and testing. The identification system shall provide for traceability of the item to its supporting documentation such as test reports or certificates of conformance.

9.0 CONTROL OF SPECIAL PROCESSES
A list of the special processes; e.g., welding, heat treatment, brazing, NDE, to be used, which may affect the quality or validity of the data or results shall be included in the project technical plan. These special processes shall be performed in accordance with approved written procedures.

10.0 INSPECTION

10.1 Checklist
An inspection checklist for the fabrication of the test article shall be developed. This may be combined with 10.2 to give a composite of the required inspections and the data that must be documented as a result of the inspection. These inspections shall include those made

after assembly and prior to any testing. All inspections shall be documented as to the results and by whom and when the inspection was performed.

10.2 Manufacturing Data

A list of the manufacturing data that will be recorded shall be part of the QA plan. These data include, but are not limited to, the following: As-built dimensions, reportable nondestructive examination indications, and surface finishes.

11.0 TEST CONTROL

11.1 Test Procedures

A written test procedure shall be prepared by the Research Organization prior to start of testing.

The procedure shall include as a minimum the following. Additional requirements may be identified in the procurement documents.

   a. Unique test procedure identification including revision level and Research Organization approvals.
   b. Data to be recorded.
   c. Method of data acquisition.
   d. Automatic controls to be used.
   e. Planned sequence of events.
   f. Range of variables to be studied.
   g. Number of tests or cycles
   h. Acceptance criteria
   i. Prerequisites.

11.2 Instrumentation

The parameters to be measured and the location, type, accuracy, and repeatability of the instrumentation shall be included in the project technical plan.

11.3 Testing Log

A log of all testing activity shall be maintained. Entries of significant events shall be made, signed, and dated on the day they occur by the testing engineer or his representative.

12.0 CONTROL OF MEASURING AND TEST EQUIPMENT

All inspection and test equipment shall be calibrated in accordance with written procedures and prescribed frequencies and shall be traceable to the National Bureau of Standards or to a natural phenomemon, when possible. If this is not possible, the basis for calibration shall be documented. At any time the accuracy of an instrument is questionable, it shall be recalibrated or not used. Any inspections or tests that were performed with equipment that is found to be out of calibration shall be evaluated to determine the validity of the inspection or test results. The results of this evaluation shall be documented. All calibrations shall be documented as to the calibration date, by whom it is calibrated, the procedure used to calibrate, and the results of the calibration. A list of the inspection and test equipment, along with their calibration frequencies, shall be maintained.

13.0 HANDLING, STORAGE, AND SHIPPING

Any special handling, storage or shipping requirements shall be identified in the project technical plan.

14.0 INSPECTION, TEST AND OPERATING STATUS

A log of testing activity shall be maintained. Entries of significant events shall be made, signed and dated on the day they occur by the engineer or his representative. If the testing takes place for an extended period, appropriate data sheets shall be maintained to indicate the status of the test at all times.

15.0 NONCONFORMING MATERIALS, PARTS OR COMPONENTS

15.1 General

Any items requiring inspection that are found to be discrepant shall be controlled to prevent usage. All nonconformances shall be reviewed by the project leader and quality assurance with the final disposition being properly documented. If the nonconformance is determined to be in violation of the requirements of the project technical plan, written approval of the disposition must be obtained from the purchaser prior to implementation of the disposition.

15.2 Testing

Any failure or malfunction of the test article shall be analyzed as to the cause and documented. The results of this analysis shall be reported to the purchaser.

16.0 CORRECTIVE ACTION

Quality deficiencies shall be reviewed to determine if corrective action needs to be taken to prevent reoccurrence. Any corrective action taken shall be documented and followed up to verify implementation.

17.0 QUALITY ASSURANCE RECORDS

The Research Organization shall establish a system for preparation, collection and retention of records sufficient to provide documentary evidence of activities affecting quality, and where applicable, of the acceptability of materials, parts or assemblies having an effect on the validity of the data or results from the data.

These records shall be consistent with applicable codes, standards, specifications, regulations and contracts and shall be adequate for use in managing the program. The records shall be identifiable and retrievable. The retention of records shall be in accordance with the contract, work order or applicable codes, standards, specifications or regulations.

The specific records to be retained and the length of retentions shall be identified for each project in the QA plan.

The final program or project report is one of the QA records. Other records shall be traceable from the project report by file numbers, references, etc.

18.0 AUDITS

18.1 Performance of Audits

The Research Organization shall have a comprehensive written system of planned and periodic audits. The audits shall verify compliance of their quality program with specification requirements and determine the effectiveness of the program. The audits shall be performed by tained personnel using written procedures and/or checklists. The audit personnel shall be familiar with all procedures and standards applicable to the areas being audited. However, the personnel performing audits shall have no direct responsibility in the area being audited. The audit shall assess the adequacy of the quality program and related written procedures. In addition, the audit shall include examination of quality operations and documentation comparison with established requirements, notification of required corrective action and timely follow-up to evaluate corrective action results.

18.2 Audit Reports and Corrective Action

The results of the audit shall be contained in a written report to the Research Organization management and made available for review by the purchaser. The audit report shall contain recommendations for correction of any reported deficiencies. The report shall be reviewed by the Research Organization management having responsibility in the area audited and action shall be taken to ensure effective correction of the reported deficiencies. Follow-up audits of the deficient areas shall be made on a frequent basis until the required corrections are made.

REFERENCES

1. Coordinating Agency for Supplier Evaluation, Aerojet Liquid Rocket Company, Sacramento, California.
2. American Society of Mechanical Engineers, New York.
3. Annual Book of ASTM Standards, American Society for Testing and Materials, Philadelphia, PA.
4. Department of Defense MIL-STD-45662, Calibration System Requirements, U.S. Government Printing Office, Washington, D.C. (1980).

Chapter 7

PROJECT CONSTRUCTION

I. MATERIAL IDENTIFICATION AND CONTROL

A discussion of the construction or erection of a test section immediately leads into the area of material control. It is in that area that the most apparent departure from standard QA techniques is observed. The term "apparent" is used because upon closer examination the differences aren't nearly as significant as one may assume. For a construction project such as the construction of a pressure vessel to an ASME code (1), the material is generally all collected into central areas and appropriately identified by tags or markings. Issuance of material from stock is closely controlled with markings properly transferred from the base stock item to the piece about to be machined or used. Often material control areas are segregated with security fences. There is documentation showing the use of the items issued on a given project or to a given person. This type of centralized stock control with limited access and controlled issuance is a vital part of material control, particularly for "code" construction. But, in a small research project, the person most likely to be involved with the material is the project leader. Since he is the one with the overall responsibility for the success of the research project, he might prefer to maintain his own controlled area for his project materials and keep them segregated from other projects. The end result is that throughout the research center there may be several storage areas maintained by the various technical sections and project leaders to assure that their materials are separate and available when the project leader wants them. Material controls, however, can still be applied to decentralized project stores with appropriate tagging and identification procedures. The project leader can maintain his identification tags and markings on the project materials and thereby demonstrate that he is appropriately controlling those materials so they are not mixed in with unidentified stock. The use of standard quality control tags, such as "hold for inspection" tags, green "accepted" tags and red "discrepancy" tags, are all available and can be used by the project leader without unduly restricting the ability to segregate special project materials from materials owned by other sections. If the project is to be constructed by a support group, such as the machine shop or welding shop, it may be prudent to establish some central storage areas. Then, when the support group has to draw some material out of general stock for a project, that material will be available and appropriately protected. This type of system is more nearly similar to the ASME Boiler and Pressure Vessel Code material control systems.

The project leader should maintain a list of materials or parts, and components that affect the validity of the test results. That list should indicate if certified test reports or certificates of conformance are required. Any additional testing by the research center to verify material identity should be specified. Copies of material certificates should be included in the project files. If the material is for a project or a test facility which is liable to be at least semi-permanent at the research center, copies of those material test reports should be on file with the QA organization. Material substitutions should be judged on the effect they have on the rest results. State and local building codes may heavily influence whether material can be substituted or not. From a purely research standpoint, the adequacy of material substitutions may be determined by the project leader. If the material was specified in the project technical plan, the project leader should obtain approval for material substitutions from the same party that approved the project technical plan.

II. CONTROL OF SPECIAL PROCESSES

One of the primary activities of an R&D Division is the development of processing techniques which may not be yet defined by any set code or standard. The refinement of an existing process may be of a nature as to advance the state-of-the-art above what is considered a production standard normally regulated by existing codes. This may be the reason for the project in the first place or it may be an ancillary process used to support the project; for example, welding of a pipeline system for experiments in flow characterizations. To the extent that special processing techniques are needed to support fabrication or for performing an experiment or test, the process technique should be documented in a formal technical procedure and approved by the project leader. Qualification of the equipment to be used should be directed by the technical procedure or by standard calibration techniques.

If personnel qualifications are required, such as for welding, appropriate qualification records should be maintained reflecting the examination results, the training records, and the maintenance of proficiencies. In those areas where the special process is to support the fabrication of a test section, existing codes may be acceptable. Those qualification processes with their documentation and testing systems developed for production work [i.e., ASME Section IX, Welding and Brazing (1)] should be quite adequate for the construction. However, if the special process is a part of the research itself, it will be up to the project leader to ensure that appropriate qualification criteria and testing methods are developed and documented. In many areas, the special process might be by laboratory personnel who are directly engaged in the research aspect of that process and who, by nature of their work, may be qualified well above standards imposed on the production facilities. These qualifications should be documented and maintained at least in the personnel department, if not by the project leader. A word of caution: there is some tendency by technical personnel to want to deviate from standardized qualification techniques or process control techniques because of their extensive experience in that field. But if the process controls or the qualification requirements are mandated by a code, then the codes must be rigidly adhered to. The temptation may be to assume that because technical sections know how to write codes, they have the capability and automatic authority to deviate from those codes. This attitude must be carefully avoided.

III. INSPECTION

Inspections are usually performed for those physical characteristics identified by the project leader as being critical to the test results. The need for inspection can be identified on receiving copies of purchase orders, route sheets, and technical procedures or inspection checklists. If desired, a special identification system can be used to denote critical dimensions on the drawings (see Figure 7.1). Critical dimensions can be determined either by the design engineer or the project leader with the assistance of QA. Dimensions so identified on the drawing would then have a corresponding checklist or route sheet to direct the inspection of that critical

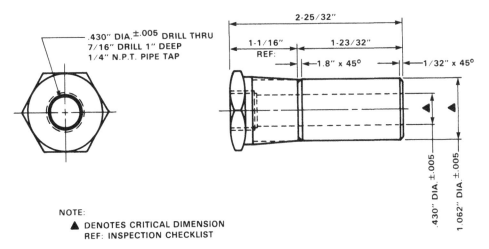

Fig. 7.1. Drawing denoting critical dimensions. (Courtesy of Babcock & Wilcox.)

dimension. In this manner, inspection efforts will be confined to those attributes that are truly important to the test results. If there are certain construction codes involving test apparatus, the inspection of the attributes that are required by the code can be handled similarly.

Inspections can be performed by any person familiar with the operation who is competent to do the operation, provided that the inspector was not responsible for performing the operation which is being inspected. Whether this should be by professional inspectors reporting to the QA organization or delegated to other organizational units with a QA overview is a matter to be determined based on the size of the project and the amount of the inspection activity involved at the research center. If compliance is required to an inspection standard or to a construction code or if the project is of a fairly large research project with construction over a period of years, then it would probably be prudent to use professional inspectors reporting to the QA organization. An alternative would be to have an inspection verification by QA. That is, critical inspections may be witnessed by the QA organization even though those inspections may actually be performed by project personnel, machine shop personnel, or someone otherwise technically competent to handle the operation. Appropriate controls over this type of activity could include a program of certification of designated inspectors. A program of training and evaluation could be required along with periodic recertification. This gives the QA organization a more direct control over the inspection function.

Another item to be considered is whether the QA organization is to be held responsible for product conformance to specified standards. An alternate would be for the QA organization to certify compliance to an overall quality plan of action. This could be verified by audit of appropriate documentation and controls. It may seem like a fine line to be drawn, but the difference is illustrated by whether the Quality Assurance Manager is asked to certify that the material conforms to specification, or to certify that the project was conducted in accordance with procedures and verified by audit? Certainly, if the Quality Assurance Manager is going to accept the responsibility for material compliance with specifications, he is going to take a more conservative approach and exercise more direct control over the inspection function. On the other hand, many inspections that are performed in the research operation are basic to the process of gathering data itself, and the project leader and his technicians have their entire laboratory set up for the sole purpose of performing these inspections and taking this data. In that event, it is the project leader and his technicians who are the most qualified to take these measurements and perform these inspections because they are going to use the results of those inspections in their calculations to arrive at some conclusion about the behavior or characteristics of that material. The QA verification process might well be limited to assuring to the project leader that the material received had proper material markings, identification and material certificates, and that some rough dimensional checks were made. The testing and detailed precision inspection would be performed by the project leader under laboratory conditions.

Occasionally, the customer will want to watch inspection or testing operations, and those "witness" points or "hold" points can be included in route sheets or technical procedures. It is then up to the QA organization to ensure that work does not proceed past any inspection points, whether customer imposed or research center imposed, without appropriate documented authorization from the customer or the R&D QA organization.

The documentation associated with inspection may vary to serve the needs of the inspector or technical section. Regardless of format, inspection results should include the identification of the inspector, the type of inspection and observation performed (such as visual inspection, dimensional inspection, or the type of nondestructive examination used), the results of the inspection, the acceptability or nonacceptability, and state directly or refer to the form that describes the action taken in connection with any deficiencies.

Some suggested formats for documenting inspections are inspection checklists, route sheets and laboratory notebooks.

Inspection checklists (see Figure 7.2) should be used if the inspection is fairly simple and if all the characteristics can be inspected at one time for a given drawing (or purchase order in the case of a receiving inspection). The receiving inspection record discussed in Chapter 6 is a specialized checklist.

Inspections of manufacturing operations (support group construction of a test article) that are more complicated are controlled by route sheets (see Figure 7.3). The receiving and fabrication inspections could be either performed by, or directly verified by, QA before the checklists or route sheets are closed out. This gives the QA organization an opportunity to verify that all dimensions were within the acceptance criteria. For dimensions outside the acceptance criteria, the appropriate corrective action forms are generated. Disposition of these discrepancies would be by the project leader.

Inspections typically performed by the technical section would be logged in the laboratory notebook (see Figure 7.4). These inspections are usually related to the condition of the test article during or after testing.

Since fabrication operations to support research are sometimes removed from the research area and located in another building under control of another organization, the gage control may not be under the direct purview of the project leader. QA can assist by reviewing the completed checklists to assure that the appropriate gages were used and their serial numbers were recorded. If, during recertification, the gage is found to be out-of-tolerance, a trace can be made back to the inspection or measurement taken by that gage.

Once the test section is delivered to the project leader, he can begin his research testing and investigations. Inspections he chooses to make can be recorded in a laboratory notebook (see Figure 7.4), along with test data

INSPECTION CHECKLIST

QA PROJECT

DRAWING NO. __9516-C__ REVISION NO. __2__ QA JOB NO. __80012__

CHARGE NO. __5168-22__ PREPARED BY __A. Kisik__ APPROVED BY __D. P. Bermingham__

ITEM	CHARACTERISTICS TO BE INSPECTED INCLUDE DIMENSIONS AND TOLERANCES, REPEAT AS NECESSARY AND IDENTIFY LOCATION	RECORD DATA	A C C E P T	R E J E C T	INSP BY	DATE INSP
1.	DRILL HOLE THRU CENTER OF PLUG TO BE .430" ±.005" DIA.					
2.	OUTSIDE DIA. OF NON-THREADED PORTION OF PLUG TO BE 1.062" ±.005"					

MEASURING DEVICE	B&W S/N	CORRECTIVE ACTION REPORT

REMARKS _____

REVIEW RESULTS   QA _____ DATE _____

PROJECT LEADER _____ DATE _____

Fig. 7.2. Inspection checklist for critical dimensions. (Courtesy of Babcock & Wilcox.)

ROUTE SHEET — PARTS LIST PAGE

DWG. NO. ARC 6524E	REV. 1	MK. NO. A-2	DATE 10/1/81	QTY 81	QA PROJECT NO. QC 81756
DESCRIPTION Plug			USED ON ASSY. Field		PROJECT NO. 10756-23 ISSUED
PROJECT TITLE Develop Plugs		PROJECT LEADER R. P. James		CAR's NR's None	

REV	DESIGN ENGR. OR PROJECT LEADER	AUTHORIZED INSPECTOR	QA J. Kimble QC	DESCRIPTION OF REV.
0	R. P. Jones	N/A	J. Kimble	Revised ARC-TP-746 Rev. 1 to Rev. 2
1	R. P. Jones	N/A		
2				
3				
4				
5				
6				

REFERENCE DOCUMENTATION REQUIRED

NO.	REV.	DESCRIPTION
QA-81756	0	
ARC-TP-162	4	
ARC-TP-746	2	
ARC-TP-54	6	
ARC-TP-340	3	

BILL OF MATERIAL

MK NO.	REV	QTY	NOMENCLATU
2	0	1	Bar Stock

ROUTE SHEET — OPERATION PAGE

DWG. NO. ARC 6524E, Revision 1	MK. NO. A-2	DATE 10/1/81	QTY 81
DESCRIPTION Plug		PROJECT NO. 10756-23	

OPER SEQ. NO	SECTION NO.		COMPLETED BY	INSPECTOR ARC, QC, AI
10	70	Perform incoming inspection and tag bar material.		J. Kimble, 10/6/81
20	52	Heat treat bar per ARC-TP-162, Revision 4.	S. Allen 10/4/81	
30	83	Machine bar per drawing 6524E, Revision 1.	J. F. Doll, 10/20/81	
40	70	QA Hold Point: Inspect and record dimensions per inspection checklist and critical dimension denoted on drawing 6524E, Revision 1.		J. Kimble 10/13/81
50	51	Assemble and test plugs per ARC-TP-746, Revision 2, and ARC-TP-54, Revision 6. Document results in lab notebook.	J. Smith	10/5/81
60	39	Perform cleanness evaluation per ARC-TP-340, Revision 3. Document results in lab notebook.	R. Shkes 11-17-81	
70	70	QA inspection.		J. Kimble 11/19/81
80	70	Release plugs for shipment.		J. Kimble 11/21/81

PAGE 2 OF 2

Fig. 7.3. Sample route sheets. (Courtesy of Babcock & Wilcox.)

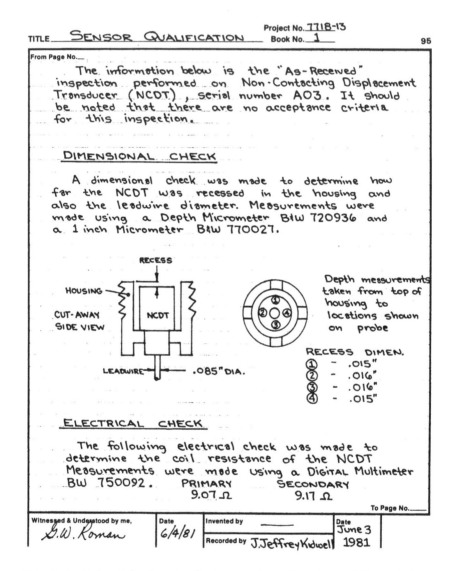

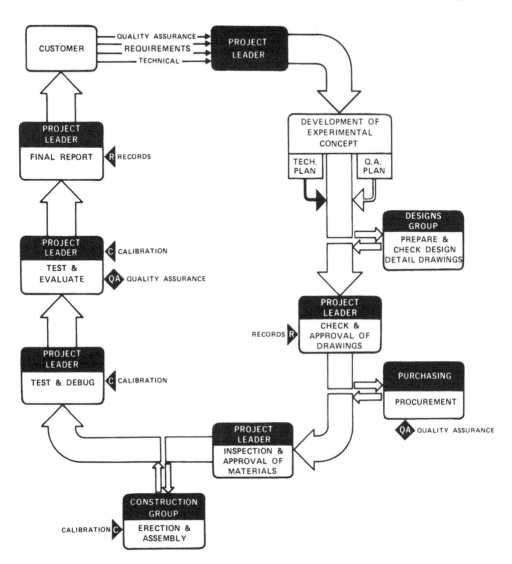

Fig. 7.5. Typical interaction of project activities with support systems. [Note: This information was previously published in Mechanical Engineering (see Ref. 2; note also the Preface).]

and a list of instruments that were used. QA can verify by audit how well
the inspections were documented in terms of who the observer was, the date
the inspection was made, and the adequacy or nonadequacy of the inspection
results.

A typical interrelationship between project activities and support systems is shown in Figure 7.5. Here the project leader has responded to the
customer's needs and developed an experimental concept which takes into
account technical and QA criteria established by the customer and technical
and QA standard systems used by the research center. Having received approval for the technical and QA plans, the project leader is, as usual, the
main impetus for accomplishing the task. In this case, the first support
system called upon is for detailed design of a test section. The technical
and QA requirements are transmitted to a design group that prepares the
test section design documentation including fabrication drawings and a bill
of materials. This design is independently checked within the designs group
for compliance with drafting standards and state and local building codes.

The design package is then routed back to the project leader. The design is compared with the technical and QA plans to assure that test section
operating conditions will be as agreed upon with the customer. At this
point, portions of the design may require an independent check if those portions were not specified by the technical plan and therefore not approved directly by the customer or if they were not reviewed within the designs
group. (Calculations made by the project leader prior to the input to designs should have been checked if they affected the designs group approach.) An "R" indicates "Records," which in this case are the retention
requirements for drawing aperture cards after the project leader approves
the drawings.

The next major support system is procurement whereby the project
leader, having determined from the test section design package which items
are "make" and which are "buy," initiates requisitions to purchasing.

The "QA" diamond indicates a supplier evaluation and approval function. The project leader then performs or arranges for inspection of incom-
ing materials. Inspection records, tags, and nonconformance reports are
standard quality forms common to any quality assurance operation — the difference is in who is doing the assurance. Whether it is more economical to
have a full-time receiving inspector depends on the individual research center.

The third major support activity is construction of the test section to
technical requirements specified on the fabrication drawings with inspection
recorded by checklists or route sheets as necessary. The calibration system is shown here for the first time supporting the project through instrument and gage control. Whether the calibration is adequate depends on the
tolerances to be measured, the capability of the gage or instrument, and
needs of the project leader. The project leader must consider these factors in selecting the proper instrument or gage as well as in reviewing the
standard calibration procedure. All measuring devices have their limitations. Conversely, it may cause unnecessary cost and delay if an exotic

measuring device is used for a simple, fairly loose-tolerance (but necessary) measurement.

After erection of the test section, the project leader may then proceed to check out the apparatus prior to taking research data. Additional calibrations may be necessary, and if a long series of tests are planned, the recall and recalibration systems will come into use. At some point, QA may elect to perform an audit to verify all quality systems are in place and effective. "Records" indicate data collection, storage, and retrieval of backup records supporting the final research report.

While the preceding example does not show all of the interactions to be expected on a given project, it does indicate the overall relationship of different systems with the main thrust by project personnel.

IV. HANDLING, STORAGE, AND SHIPPING

Special handling, storage or shipping, packaging, preservation or cleaning instructions required for test articles should be identified by the project leader in the contract proposal or the project technical plan. The project leader should either issue a technical procedure detailing these requirements or coordinate with the design engineer to have the requirements defined by drawing. The technical procedures would include provisions for recording inspection of the work, if such inspection is required. If drawings are used, the associated route sheets and inspection checklists document accomplishment of the action and the inspection.

V. NONCONFORMING MATERIALS, PARTS, OR COMPONENTS

Nonconformances detected during the fabrication of test articles should be identified on an appropriate nonconformance report. The article itself should be tagged with some significant identifier, such as a red discrepancy tag, and segregated whenever possible. The project leader should then review the nonconformance to determine any adverse effect to the test results. The project leader should approve actions taken to correct the nonconformance or choose to use the article as is, with QA exercising approval over the corrective actions. When the nonconformance is satisfactorily resolved, the discrepancy tag can be removed from the article. In instances where a "repair" or "use as is" decision is contemplated, the project leader must determine whether such a decision will cause a deviation to any of the customer requirements or any commitments made in the proposal or project technical plan. If a deviation will exist, the project leader has the responsibility to obtain approval from the customer before he approves the repair or use as is decision. Obviously, nonconformances requiring supplier corrective action should be coordinated with Purchasing.

QA should take direct action for nonconforming items by verifying that the appropriate corrective action has been implemented to fix the nonconformance and to assess whether long-term corrective action is warranted, either from the manufacturing standpoint or from the quality systems stand-point. This also puts QA in the appropriate information loop to gather de-fect data and perform long-term collective analysis of various research projects so undesirable trends or chronic violations can be detected and corrected. QA should always be involved in the corrective action process so that it may exercise its particular skills in the problem solving aspects of the quality program. An example of an analysis of corrective actions is given in Appendix "A" and shows how a simple statistical tool can be quite effective.

The Corrective Action Report shown in Figure 7.6 has a minimum of check type entries. It is a flexible form with which the initiator is given a wide latitude in reporting the problem. It may be desirable to separate manufacturing defects from system defects to permit more specialized and rapid data collection and analysis. Separate discrepancy reports can be designed to more closely suit the process. The basics of recording the discrepancy, identifying the action to be taken and follow-up to verify im-plementation apply to any form used.

The organization or person with primary responsibility for the QA program must be in a position to determine when an incident or condition has occurred which could jeopardize the attainment of quality objectives. This is so that an appropriate report can be made to the cognizant project leader, group supervisor, or upper level management. After the recipient has conducted an investigation for the cause of the incident or condition and documented the results of the investigation, QA can then follow-up within a reasonable time to assure that whatever corrective action has been agreed upon, has been implemented, and is effective. Whenever any quality sys-tem deficiency is considered to be significant, QA should have the preroga-tive to require a work stoppage in the effective area until the condition has been corrected. With the prerogative, QA may not always assume the role as primary investigator and determiner of the corrective action, but is in the position to require that an appropriate investigation is done by compe-tent persons and that an agreeable solution has been obtained.

APPENDIX: AN ANALYSIS OF CORRECTIVE ACTION REPORTS
GENERATED DURING A 3-YEAR PERIOD

An analysis was performed of the Corrective Action Reports generated at the research center for a period of three and a half years (3). The purpose was to show which administrative procedures were being violated. This would indicate areas where project leaders were having the most problems. The results would indicate which procedures were not being properly com-municated or understood. The initial analysis might direct attention to areas requiring a more intensive audit or investigation for program correc-tive action.

QUALITY ASSURANCE   NUMBER   C-824

RESEARCH & DEVELOPMENT DIV'   **CORRECTIVE ACTION REPORT**

PROJECT Effect of XYZ Lubricant on the Stress Corrosion Cracking of AISI Stud Material	JOB NUMBER QA-81022, 5291-01

PROBLEM DESCRIPTION

Entries in the lab notebook show that the test system was not maintained at operating conditions for a minimum of 14 hours as specified in the Project Technical Plan. This problem is applicable to all five tests performed.

INITIATOR S. L. Molnar	QA MANAGER G. W. Roberts	DATE 3/3/82

TO NAME John V. Doe	SECTION Corrosion Technology	DATE REPLY DUE 3/15/82

RESULTS OF INVESTIGATION AND/OR CORRECTIVE ACTION

Our normal laboratory test procedure is to let the test system stabilize overnight or a minimum of 14 hours before applying a load to the specimen. Due to the customer's request to perform the test series as quickly as possible, this was not followed. However, this will not affect the results obtained for the five tests performed.

The actual time at temperature before the load was applied is recorded in the laboratory notebook for each test.

A revision will be made to the Project Technical Plan.

SIGNATURE John V. Doe	DATE 3/3/82

TO: QA MANAGER

QUALITY ASSURANCE REVIEW OF PROPOSED CORRECTIVE ACTION	☒ SATISFACTORY ⎯ UNSATISFACTORY	SIGNATURE	DATE 3/4/82
QA FOLLOW-UP TO VERIFY IMPLE- MENTATION OF CORRECTIVE ACTION	☒ SATISFACTORY ⎯ UNSATISFACTORY		

The revised Project Technical Plan was approved by the customer.

	QA SIGNATURE	CLOSED OUT DATE 3/4/82

Fig. 7.6. Corrective action report. (Courtesy of Babock & Wilcox.)

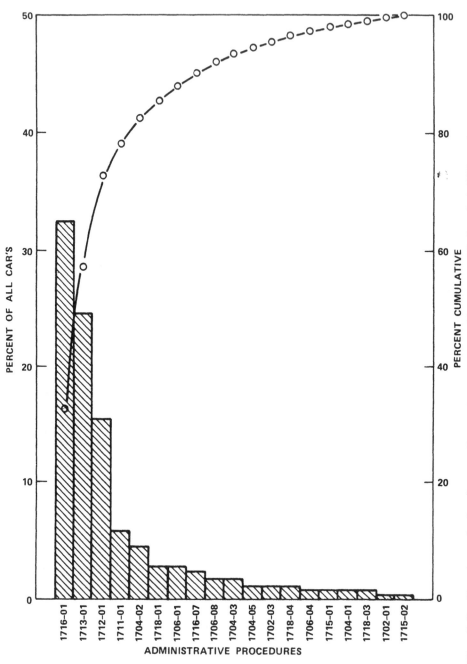

Fig. 7.7. Pareto analysis of corrective action reports. (Courtesy of Babcock & Wilcox.)

Using basic Pareto analysis techniques described by Juran (4), the graph in Figure 7.7 was developed. It displays evidence that the five procedures on the bottom left accounted for more than 82% of Corrective Action Reports for the 3-1/2 year period. These procedures covered the following subjects:

1716-01	Hardware Defects
1713-01	Calibration System
1712-01	Use of Written Test Procedure
1711-01	Use of Inspection Checklists
1704-02	Preparing Project Technical Plans

The results suggested that most of the problems (32.5%) were due to hardware defects — a fact not unexpected since the Corrective Action Report is used to document defects in fabrication of test articles as well as programmatic violations. It did, however, point out a need for a more thorough system for collecting defect data. A separate nonconformance report could have been more suitably designed for classifying hardware defects.

The second highest contributor (24.5%) was in the calibration and control of test equipment. For a research center, this might also be expected, but definitely undesirable. An intensive program of system auditint was established using sampling techniques for each category of instrument. Significant progress was made in reducing the errors in this system.

The remaining three administrative procedures were associated with proper initiation and use of documents associated with the projects. These problems were treated by revision to the procedures and associated forms to clarify instructions and by working with project leaders and designated inspectors to help them better understand the systems.

Although this case history did not involve rigorous statistical investigation techniques, it did demonstrate a useful means for evaluating administrative systems.

REFERENCES

1. American National Standards Institute, ANSI/ASME BPV-1, ASME Boiler and Pressure Vessel Code, The American Society of Mechanical Engineers, New York (1981).
2. G. Roberts, Quality Assurance in R&D, Mechanical Engineering, Vol. 100, No. 9, p. 41 (1978).
3. J. Black, Internal Report, Babcock & Wilcox, Alliance, Ohio (1978).
4. J. Juran, Quality Control Handbook, McGraw-Hill, New York (1974).

Chapter 8

CONTROL OF
MEASURING AND TEST EQUIPMENT

I. THE NEED FOR FORMAL CONTROL

A measurement control program is one of the most critical programs within a research center. Marguglio states, "Probably the most pressing measurement control problem existing in industry today is the failure to exercise adequate control consistently over measuring devices which are being used for design development, for example, for developmental test (1)."

Dr. B. C. Belanger, Chief of the National Bureau of Standards (NBS), Office of Measurement Services, has observed that

> scientists and engineers who publish data in archival journals should be concerned about making the experimental data as accurate as possible so that other scientists and engineers can use them with confidence to duplicate it. Most scientists do not document measurement traceability to National Standards, since there is little incentive for them to do so. By contrast, a Military Contractor who faces economic consequences for not meeting traceability requirements will act accordingly. In view of the importance of measurement accuracy in science and technology, it is surprising how few scientists take the time and effort to systematically develop an uncertainty statement for their data (2).

There are government standards available which provide system requirements for a measurement control program. Most notable, of course, is MIL-C-45662A (3) and its successor MIL-STD-45662 (4). These concentrate on programmatic controls to assure proper periodic comparisons of standards to assure "traceability" to NBS. In addition, there is also a great deal of important work currently underway to formulate a national consensus standard embodying the latest concepts in measurement assurance. This work includes controls for developing the uncertainty statement mentioned by Dr. Belanger.

II. TRACEABILITY

Before embarking upon a discussion of what types of actions are recommended to control measuring and test equipment and to ensure "traceability," it would be well to discuss what traceability is and is not. While re-
searching for this chapter, the author came across a very comprehensive
work by Dr. B. C. Belanger (2) of NBS. Part of the work has already been
quoted above, but significant sections are reprinted below and on the next
five pages.

A. Definition of Traceability

Dr. Belanger cites four possible definitions of traceability without indicating
his preference:

1. Traceability is the ability to demonstrate conclusively that a particular
    instrument or artifact standard either has been calibrated by NBS at ac-
    cepted intervals or has been calibrated against another standard in a
    chain or echelon of calibrations ultimately leading to a calibration per-
    formed by NBS.
2. Traceability to designated standards (national, international, or well-
    characterized reference standards based upon fundamental constants of
    nature) is an attribute of some measurements. Measurements have
    traceability to the designated standards if and only if scientifically rig-
    orous evidence is produced on a continuing basis to show that the meas-
    urement process is producing measurement results (data) for which the
    total measurement uncertainty relative to national or other designated
    standards is quantified.
3. Traceability means the ability to relate individual measurement results
    to national standards or nationally accepted measurement systems
    through an unbroken chain of comparisons.
4. Traceability implies a capability to quantitatively express the results of
    a measurement in terms of units that are realized on the basis of ac-
    cepted reference standards, usually national standards.

    While it is apparent that there are similarities in these alternative
definitions for the apparently elusive concept of traceability, it is perhaps
more interesting to note the inherent differences and their implications.
Definition 1 focuses on an unbroken chain of calibrations of instruments or
standards, with the chain ending at NBS, while Definition 2 does not even
mention calibrations explicitly. A comparison of Definition 1 with Definition 2 provides perhaps the clearest exposition of two contrasting views of
traceability, the first stressing characteristics of measuring instruments
or standards while the second stresses requirements related to quantifying
measurement uncertainty. The one regards accuracy as a property of an
instrument, while the other focuses on the quality of the measurements
themselves. The implications of these two points of view should become
clearer as this discussion continues.

It is important to stress at the outset that the resolution of the issue of whether or not a particular system for realizing traceability is effective depends entirely on what one assumes is the real <u>intent</u> of traceability. This author believes the real intent of traceability is to insure measurements of adequate accuracy.

B. Purposes and Uses of Traceability

It is logical to assume that any individual or organization intending to purchase a product or service that involves measurements by the supplier organization would want to have the supplier take steps to ensure that the measurements were sufficiently accurate for the product or service to meet the customer's needs. Procurement contracts do sometimes contain clauses that require the supplier to "maintain traceability to national standards," but too often such contracts do not explicitly define traceability or specify what constitutes adequate traceability. One cannot find fault with the intent of such citations, but one can legitimately raise the question of whether or not all parties have a consistent interpretation of what constitutes an adequate methodology for establishing and maintaining traceability.

A very well-known and important example of traceability requirements can be found in DOD procurement policy. For many years, DOD has written clauses into its procurement contracts that require the contractor to be able to demonstrate traceability (1). The most familiar example of this is MIL-C-45662A, Calibration Systems Requirements, which states in Section 3.2.5.1:

> Measuring and test equipment shall be calibrated by the contractor or a commercial facility utilizing reference standards (or interim standards) whose calibration is certified as being traceable to the National Bureau of Standards, has been derived from accepted values of natural physical constants, or has been derived by the ratio type of self-calibration techniques.

Disputes between DOD and its suppliers over what this section means are infrequent, even though this well-known specification does not explicitly define traceability. Section 3.2.3 states:

> Measuring and test equipment and measurement standards shall be calibrated at periodic intervals.

This uncertainty indicates that calibration is a key element in establishing traceability in the opinion of the authors of MIL-C-45662A.

It is easy to see how a contractor attempting to comply with this specification would tend to gravitate towards a system in which a calibration echelon, established recalibration intervals, and documentation of calibration records would be emphasized. Such a traditional approach may be a necessary condition for traceability, but may or may not be sufficient, depending on the intent of the requirement. It is certainly possible for a facility to make inaccurate measurements even though its standards have been recently calibrated by NBS.

C. Measurement Traceability and Standards Traceability

If one adopts the first of the four definitions of traceability given above, it can be shown that traceability by itself is not an adequate demonstration of measurement quality assurance. It may, however, be an important component of a measurement quality assurance system, If one adopts the second definition, then traceability is much closer to being synonymous with measurement quality assurance. Traceability has traditionally — for example, in compliance with MIL-C-45662A — focused on a calibration chain for instruments or standards. That is, the language used implies that instruments or artifact standards must be traceable to NBS. In the opinion of many persons at NBS and elsewhere this will not generally ensure acceptable accuracy unless additional steps are taken since having an accurately calibrated instrument is no guarantee that a measurement made with it will also be suitably accurate. Operator errors, unusual environmental conditions, and the like can drastically affect individual measurements and, thus overall accuracy. In real life, one cannot always afford to have highly skilled metrologists; errors may thus go undetected. DOD procurement requirements have moved more and more in recent years toward an approach in which calibration is integrated into an overall quality assurance system.

D. The Question Posed to NBS

NBS is often asked questions such as "If I have my device calibrated by A and A has his device calibrated by B and B has his device calibrated by NBS, have I achieved traceability?" By now the reader should have enough insight to appreciate why such questions are difficult to answer. The answer in a legalistic sense hinges on another practical question: "Is your auditor or contract monitor satisfied with your measurement capability?" Traceability, like beauty, may be in the eye of the beholder. When one adopts a definition like the second one given at the beginning of this article, an important question is, "Can you demonstrate that the uncertainty of your measurements is sufficiently well known and sufficiently small to meet all of your requirements?" If the answer is "no," or "I don't know," then corrective action should be taken.

The total uncertainty associated with any measurement consists of the random error (variability within the laboratory) and the systematic error (bias or offset) of that laboratory relative to national or other standards. Most traditional approaches to traceability, whether they involve calibrations SRM's, tend to focus only on the systematic error. Since total measurement uncertainty generally determines whether or not the measurements are adequate, it should be the basis for a rigorous approach to measurement quality assurance. Thus, requirements that specify measurement uncertainty explicitly are preferable to requirements that only call for traceability to NBS implying the traditional interpretation of traceability.

An occasional but real problem for the metrology community is illustrated by the hypothetical example of a manufacturing firm that claims traceability to NBS for its products, but doesn't tell the customers that the traceability claim is based on one calibration of an artifact by NBS in 1932, and that the artifact has been lost for the past 15 years. NBS has no police powers to redress such abuses (nor do we desire to be given such powers). The Federal Trade Commission or Department of Justice might well hesitate to prosecute such companies for false advertising since the meaning of the term traceability is obviously subject to varying interpretations.

The question is sometimes asked, "If I have accurate measurements relative to national standards but can't prove it to anyone, do I have traceability?" By Definition 2, if you have traceability, you automatically have the means to prove it to the satisfaction of any knowledgeable technical person.

E. NBS Advice on Traceability

> As requirements for traceability to NBS proliferate, NBS is being deluged with inquiries from individuals and organizations demanding to know: What must I do to be traceable to NBS to satisfy the requirements of this or that regulation or the terms of this or that contract.

Often the questions begins, "What is the minimum I must do ... ") When such requests are received, NBS tries to be helpful by providing references to technical papers describing good measurement practices and information on measurement services provided by NBS, other government organizations, and private firms. We advise the person making the inquiry that NBS cannot legally say whether or not a given practice is adequate since this must ultimately be the auditor or inspector's decision. If a regulator and someone being regulated or DOD and a military contractor disagree on a compliance issue relating to measurement, NBS may be willing to mediate if both parties agree. However, NBS maintains that final determination of compliance or noncompliance is not legally our responsibility. Nevertheless, we do believe that it is important that NBS be involved in discussions within the metrology community and the voluntary standards community concerning the meaning of traceability and its achievement.

One key function of NBS is to maintain the basic SI units and to provide access to these and many derived units of interest to the technical community through a variety of measurement services. Occasionally someone will require traceability to NBS for some measurement for which we do not maintain standards or provide measurement services. When this occurs, it may create considerable confusion until all concerned parties recognize that direct traceability to national standards cannot be achieved where national standards do not exist. Since there are literally thousands of derived units and an even larger number of measuring instruments and standards associated with those units, at any given time NBS can only provide calibration services for a selected number of items. The services offered by NBS are then deemed most necessary to support current technology. As technology changes, NBS services also change, and NBS often initiates new calibration services and terminates those that are seldom used.

In most cases where it is necessary to achieve traceability to national standards for a highly derived unit, it is possible to do so by a "boot strap" or indirect process. For example, NBS does not calibrate speedometers, but our services provide access to the meter and the second so that the user can in principle accurately measure velocity by employing good metrology practices. By quantifying the uncertainty of velocity measurements in a scientifically rigorous manner and demonstrating their accuracy on a continuing basis, traceability (as per Definition 2) can be achieved in spite of the unavailability of NBS speedometer calibrations.

One can also apply the discussion in this article to issues of traceability to international standards. By international agreement, the artifact kilogram kept at the International Bureau of Weights and Measures (BIPM) near Paris, France, is defined as exactly one kilogram. Since the United States and most other large countries have nearly identical copies of this kilogram, measurements of mass around the world are compatible and it can be said that they have traceability to the international standard for mass.

The issue becomes more complex for derived quantities like mechanical vibration or microwave power. For such measurements there is no international analog to the kilogram in France. From time to time international intercomparisons for certain derived units to take place on a bilateral basis; where this has occurred, one has an estimate of the possible systematic errors between countries. There is at present no comprehensive worldwide system for insuring measurement uniformity. One cannot today go to any country in the world, have an item calibrated by its national standards laboratory, and thereby determine its uncertainty relative to U.S. National Standards. NBS is now exploring with other national laboratories, on a limited basis, what would have to be done (both politically and technically) to achieve international reciprocal recognition of calibrations or other measurement services.

It is clear that NBS cannot calibrate every measuring instrument and standard used in the United States. The capability for accurate measurements and calibrations must be distributed throughout the country among the private sector and government organizations that need it. For instance, at one time NBS calibrated nearly every fever thermometer sold in the United States. When the state-of-the-art for temperature measurement in U.S. industry had progressed to the point where it was clear that industry had the capability to protect the public safety by manufacturing thermometers that were adequate for their intended application, and when suitable consensus standards became available, NBS phased out such calibrations. We now limit our calibrations of thermometers to those standard thermometers needed to verify the continuing accuracy of temperature measurement in domestic standards laboratories.

NBS intentionally limits its services, first, to those special measurement services at the highest levels of accuracy that provide a means to quantify bias relative to national standards, and second, to services where NBS has some specialized expertise that does not exist elsewhere. We make every attempt to avoid competing with private sector calibration laboratories by duplicating their offerings.

A challenging area for future NBS attention is the subject of traceability for automated test equipment (ATE) and dynamic measurements. For complex computer-based equipment, it is not enough to insure that the built-in reference standards are accurately calibrated at regular intervals. If the adequacy of the software and interfaces is not completely verified, the system may be producing inaccurate measurements of the unit even though documentation can be produced to show that the standards incorporated in the ATE system are "traceable." We at NBS are working closely with the Industry/Joint Services Automated Test Project and organizations like the Calibration Coordination Group in the DOD and the National Conference of Standards Laboratories to address this issue now. There seems to be a consensus that the surface has hardly been scratched.

The concept of traceability seems to be evolving in the direction of measurement as opposed to standards traceability, and this is as it should be. The author has discussed the concept of traceability with many of the NBS staff and a number of knowledgeable people outside NBS, and has found a strong consensus that something along the lines of Definition 2 is much preferable over the other possibilities. Definitions 3 and 4 may be considered to occupy an intermediate position between Definitions 1 and 2.

III. PROGRAM DESCRIPTION

Any major program should be described in writing. People want and need to understand what the system does, who is responsible for what specific actions, how and when to enter the system, and what to expect from the system. As one of the programs critical to the success of research and development (R&D), the measurement control system should be formally documented to clearly define the responsibilities of all technical and support sections using measuring devices at the facility.

The program description should be sufficiently detailed to permit periodic audit by an independent agency. Technical specialists sometimes get too involved in the technical details of their project and consequently calibrate their instruments just enough to meet the specific ranges they are concerned with. They tend to cut corners in areas they feel are of least importance to their current project. They are even less concerned with establishing a long-term history of the instrument. As a result, the documentation tends to be incomplete. Periodic audits will emphasize proper maintenance of instrument documentation and examine the calibration program from the perspective of benefit to all projects, not just one.

IV. SCOPE

The program should specifically include all measuring devices used at the facility. Some means of physical identification of each device is needed to show that the device was incorporated into the system. By examining the unit, one can immediately tell that appropriate files have been established. Therefore, the unit may reasonably be assumed to be under control. Lack of such identification leaves one ignorant of who the owner is, when it was bought or built, and what its condition is. One may assume the unit was bought from a local discount store (which it may have been), instead of a reputable instrument manufacturer — hardly an item to stake your company's reputation on.

The identification (usually serializing) does not have to imply that all units are calibrated or even capable of being calibrated. That message is given by the calibration label. The serial number does state that an evaluation was made upon receiving the unit at the facility and that appropriate decisions were made with regard to the unit's ownership, location of use, history files, and planned periodic maintenance. Manufacturers' serial numbers cannot be used for this since they obviously were attached before such an evaluation could be made.

V. QUALITY LEVELS

As with quality program planning, the preassignment of quality levels may be useful for measurement control. After evaluating the unit's design and intended use, it might be desirable to assign the unit to a category of instruments for which a high degree of confidence is required. This category might include primary and transfer standards and those working level or product measurement instruments used for inspection or to take test data. These instruments would be calibrated just prior to and just after use or be on a specific time interval for recall, recalibration and maintenance. The calibration process would be tightly controlled with respect to operator, procedure, environment, and reference standards to ensure that an accurate statement of uncertainty can be made with respect to a natural phenomenon or an artifact maintained by the U.S. National Bureau of Standards or other recognized National Standards agencies.

A second category could be established for those devices which by design are able to be calibrated, but by reason of intended use it may not be desirable to do so, or at least not on a regular basis. This type of instrument would be used in a noncritical application for reference use only, while actual data was being taken by a Category 1 type of instrument. If this class of unit is used, it should be clearly identified to avoid any confusion with a Category 1 unit.

A third category might be useful for units which, by design, cannot be calibrated. They cannot be used for inspection or to take data, but they are used in conjunction with a Category 1 or 2 unit. For these items, a functional check or some periodic maintenance is desirable. An example of this is a power supply.

The classification of units into categories based upon intended use provides an effective means to preplan costs of maintenance without jeopardizing quality. Each of the categories are auditable to a basic set of system controls established for that category. Audit sampling plans can be adjusted to suit the degree of confidence required for each category. Each category should be clearly labeled, and movement from one category to another should require extreme care. For example, before downgrading from Category 1 to Category 2, a calibration should be performed using the Category 1 criteria. This will assure that test data taken subsequent to the last Category 1 calibration did not suffer from instrument drift or if the readings did drift, an appropriate correction factor can be applied to the test data. After taking care to assure prior data is not affected, the unit may then be downgraded to a lower category.

VI. CENTRALIZED CONTROL

Technical specialists may argue that for "their" particular field they have no need for a centralized structured program to meet a specification such as MIL-STD-45662 because in their area they can show technically how they assure their measurements are good. The chemist will show use of pure, laboratory grade standard reference materials for a spectrographic analysis; since his reference is an element, the comparison is to a natural phenomenon. Rigid procedures may be in place for comparing mass standards periodically to provide statistical evidence that nothing has changed; therefore, there is no need to send reference standards back to NBS. Others may argue that comparing a working standard to a reference standard that is 10 times more accurate just to verify tolerances without taking any data is meaningless; therefore, complying with a standard that seemingly allows this to be acceptable is worse than useless — it is misleading. Finally, there are those who do not wish to have their equipment handled, transported or adjusted in any way by anyone but their own people. Since their own people must risk their reputations on their data, the fewer people tinkering with the mechanisms, the better. All of these arguments can and will come up in any research lab with its variety of measuring functions. All of them may be used in an attempt to resist the imposition of a centralized measurement control system. This is particularly so if there is a feeling that a military standard was selected just because it was the nearest applicable standard and "if the government does it, it must be good. " The application of any specification must be shown to have merit, and the detailed implementing procedures must permit variations within the research lab.

It is not mandatory that a gage room be under administrative control of a QC or QA department for the sole purpose of complying with MIL-C-45662A, and it is not mandatory that all items in the program belong to the gage room and be charged out by using sections. These are actions taken by some companies, and which, for them, was the best way to meet their needs. They were company decisions, not a requirement of military standards.

A centralized control concept can be effectively initiated by placing the authority for "certifying" the calibration status within one group having technical competence to oversee the calibration function. This group would have the power to perform the initial evaluation and assignment of categories, approve procedures, test setups, test personnel qualification, and test data (whether generated by its own people, other technical sections, or suppliers). After evaluation of the data, the central control group can either accept or reject the data. If the data is rejected, the using group must be notified so it may make its own evaluation of the impact of an out-of-calibration condition on previous experimental test results. If the data is acceptable, the "certification" group can issue an appropriate certification label for the item and update the calibration status list.

The central control group's operation should be subject to overview and audit by QA, as are the other technical groups at the center. This can be done as part of a project audit or separately as a system audit (see Chapter 12).

With all of the above controls in place, it remains the responsibility of the user to assure acceptability of the instrument for the data to be taken. This includes being aware of each unit's ranges, accuracy, and repair history and how they affect the overall accuracy and repeatability of the total measurement process.

VII. CALIBRATION PROCEDURES

The calibration process must be in a state of control to reduce the degree of variability. Written calibration procedures are an aid to establishing this control. Manufacturers often provide calibration instructions; these should be reviewed to make sure they are adequate and appropriate for the user. Test bench equipment may vary, and the type of equipment to be used should be compared with the manufacturer's recommended setup to assure consistency of results. Often, manufacturers' recommended procedures do not adequately address the user's needs, and specialized procedures are necessary. Even after standardized procedures have been developed by the research center, the occasion may arise when alternate procedures have to be developed by a research group to suit their specific needs. All procedures, whether supplied by the manufacturer, a central control organization, or a research group, should be reviewed and approved by the central control agency. This review will determine if the alternate procedure is an improvement that should be incorporated for general use or whether it is a limited use procedure. Also, the effect on the calibration status of the unit can be determined.

Some guidelines which may be helpful in writing calibration procedures:

a. List literature that normally would be required during the calibration. This should include such items as manufacturers' instruction manuals and conversion tables.

b. Identify all equipment needed during the calibration. List the general requirements of the equipment first and then a recommended device that will satisfy those requirements. Do not specify actual serial numbers of the equipment in the procedure — put that information in the instrument service log (see Figure 8.1).

c. Define tasks of a general nature which need to be performed before the actual calibration can be started. These would include things like preparation and setup of standards, measurement techniques, etc.

d. If the calibration setup is not straightforward, a hook-up diagram should be included to ensure uniformity of calibrations.

e. Provisions should be included to check the "as-received" or initial calibration condition of the instrument before repairs or adjustments are made. Criteria should be included to allow the person doing the calibration to determine if the unit is within tolerance limits. A set of data should be taken per paragraph f below and recorded for this test. If the unit is out of tolerance, the user should be notified and provided the data. An evaluation can be made by the user of the impact of the data shift on previous test results (see Figure 8.2).

f. The test section should list the steps necessary for the calibration of the device. If the readings taken during the precalibration check (paragraph c above) were within tolerance and no adjustments were made during maintenance, then it may be possible to make a quantifying statement that would eliminate the need for an additional calibration. Include such things as data points to be measured, tolerances, special techniques, and acceptance criteria. A sample data sheet can be included for complex tests to ensure consistent recording of data.

g. Include a description of any maintenance or cleaning required before returning the unit for use.

INSTRUMENT SERVICE LOG

MANUFACTURER _____	SERIAL NO. ☐☐☐☐☐☐	
MODEL NO. _____	MFG. SERIAL NO. _____	
INSTRUMENT _____	GOVERNMENT NO. _____	
REQUESTED BY _____	CHARGE NUMBER _____	
SECTION _____	WORK ORDER NO. _____	

Location _____ Category (1, 2 or 3) Manhours   PROCEDURE NO. _____ CERTIFIED FROM _____ TO _____

☐ ☐☐

Timer Reading	User Group	Material	Date of Service Mo. Day Yr.	Recall Interval
☐☐☐☐☐☐	☐☐☐	☐☐	☐☐ ☐☐ ☐☐	☐☐

STANDARDS USED

Mfg. and Item	Model No.	Ser. No.

REPAIR PARTS	COST	AUXILIARY EQUIPMENT USED	Ser. No.

Use Other Side, or Attach Separate Sheet Showing Calibration Curves and Special Test Equipment Setup.

Problem Description & Service Notes	Calibration Data

Name	Date

Fig. 8.1. Instrument service log. (Courtesy of Babcock & Wilcox.)

TO: _____ *(Equipment User)* _____

OUT OF CALIBRATION REPORT

INSTRUMENT_____ *(Mfg., Model, Item)* _____   SERIAL NO. _____

WAS FOUND TO BE OUT OF CALIBRATION ON _____ *(Date)* _____

THE FOLLOWING READINGS WERE OBSERVED. _____

_____

_____

(Calibration Readings)

_____

_____

_____

_____

_____

_____

_____

_____

_____

_____

_____

_____

THE LAST PREVIOUS CALIBRATION WAS ON _____ *(Date)* _____

YOU MAY USE THIS INFORMATION TO EVALUATE POSSIBLE IMPACT ON PREVIOUS TEST RESULTS.

BY: _____ *(Person Who Did the Calibration)* _____   DATE: _____

THIS PORTION IS TO BE FILLED OUT BY THE EQUIPMENT USER. PLEASE REPLY WITHIN 30 DAYS.

I HAVE EVALUATED THIS INFORMATION FOR ITS IMPACT ON THE DATA COLLECTED FOR PROJECT NO. *(List All Order Numbers Involved)* THIS EVALUATION HAS BEEN INCLUDED IN THE PROJECT FILE.

BY: _____ *(Equipment User)* _____   DATE: _____

Fig. 8.2. Out of calibration report. (Courtesy of Babcock & Wilcox.)

VIII. HISTORY FILE

Each unit should have a file to collect calibration data and repair history (see Figure 8.1). The file should be started as soon as the instrument is received and should include procurement data. As experience is gained from using the equipment, recalibration intervals can be adjusted and preventive maintenance procedures upgraded.

IX. EVALUATION OF CALIBRATION SUPPLIERS

A company may elect to accept calibration certificates at face value just as they would certificates of conformance for material or manufactured items. The same dangers are inherent in either situation. A precept of quality is that a "vital few" contribute to most of the problems. In R&D, the area of procurement most vulnerable to supplier variations is calibration data, since that is the basis upon which R&D data is founded.

A list of qualified outside suppliers should be developed for those instruments that cannot be calibrated at the research center. A program of supplier evaluation, including periodic audit such as that described in Chapter 12, will help provide confidence that a supplier can back up his certifications. The time spent with the supplier will help him to understand your particular needs. In return, you will obtain a better understanding of the supplier's standard system and assess the degree to which he can react to your needs. The more he has to deviate from his standard system to accommodate your order, the more likely an error will be made in handling your order.

If the supplier is to be used for recalibration, the procurement order should specify a precalibration check to be run, as described in paragraph VI.e, with the data reported back to the research center so that appropriate corrections can be factored into previous data.

REFERENCES

1. B. W. Marguglio, Quality Systems in the Nuclear Industry, Special Technical Publication 616, American Society for Testing and Materials, Philadelphia (1977).
2. Dr. B. C. Belanger, Traceability — An Evolving Concept, ASTM Standardization News, Vol. 8, No. 1, p. 22 (January 1980). Exerpts reprinted by permission of American Society for Testing and Materials, Philadelphia, PA.
3. Department of Defense MIL-C-45662A, Military Specification, Calibration System Requirements, Washington, D.C. (1980).
4. Department of Defense MIL-STD-45662, Military Standard Calibration Systems Requirements, Washington, D.C. (1980).

I. TEST PLAN

ANSI/ASME NQA-1 (1) includes a requirement for test control which states:

> Tests required to verify conformance of an item to speci-
> fied requirements and to demonstrate that items will per-
> form satisfactorily in service shall be planned and executed.
> Characteristics to be tested and test methods to be employed
> shall be specified. Test results shall be documented and
> their conformance with acceptance criteria shall be evalu-
> ated.

Research centers are in the peculiar position of being the user of
their own fabricated units. Test controls are applied not only to the tests
to determine their acceptability of an article before being placed into use
(as addressed by the above ANSI requirement), but also to those tests performed with the unit in service to gather data about that unit or another
phenomenon.

Tests should be planned beforehand to establish an agreement as to
what is to be studied and what characteristics are to be varied. The overall
test plan should be included in the project technical plan and is based on the
design of the experiment to be performed. Design reviews, unfortunately,
concentrate on physical design of the test facility, the prototype test article
or the production hardware. The design of the experiment does not always
benefit from the same rigor of a formal review, but if the overall test program is planned and documented in a technical plan and subjected to approval
by the purchasing and performing parties, there is at least an agreement
from a contractual standpoint. Whether the test program is statistically
sound, rather than merely financially acceptable, may have to be determined by an additional review of the technical plan. This should be done
prior to actually embarking upon an expensive test program. Requirements
for customer or third party witness of the tests should be included in the
test plan, along with notification agreements.

II. TEST PROCEDURES

Once the overall test plan has been agreed upon (by approval of the project technical plan), the construction plans for the test article can proceed. Control of specific tests described by the overall plan is exercised by individual test procedures. Many of these are standardized tests, such as those published by the American Society for Testing and Materials (ASTM) (2), and the American Society of Mechanical Engineers (ASME) (3). These can be identified in the project files by number and revision, since they are readily available for future reference. Other procedures must be written especially for each test. Some may be very open-ended and must be developed as the test proceeds. The usual practice in this case would be to write a test procedure with only those steps and acceptance criteria that are known beforehand. The idea is to specify as much of the detail of the test as is known at the time, then amend the procedure as necessary to add testing. Obviously, in a situation where the phenomena are being characterized, there is no acceptance criteria. This is not a performance or qualification test to be passed or failed. Therefore, the procedures should only require the test data to be recorded in some prescribed format on data sheets or in a laboratory notebook.

Detailed test procedures should identify prerequisites, such as special operator training or certification requirements and test conditions that should be verified before the test begins (see Figure 9.1). If tests are sequenced, the procedure should require verification of completion of prior tests before proceeding with the current tests. If there is a long series of tests to be performed to different test procedures, a route sheet is beneficial for directing progression through the testing process, much as a shop route sheet (traveler) directs manufacturing processes (Figure 7.3).

A listing of typical test equipment to be used with the required range and accuracy should be included in the procedure, but the listing of actual serial numbers should be avoided since if units must be changed for any reason, the procedure must also be revised. It is important, however, to know what specific units were actually used during the test. A block diagram (see Figure 9.2) of the test setup is very useful. This information can be logged on data sheets or in the lab notebooks. If duplicate units are used, it is important to specify which serial numbers were used in which locations. If, during the recalibration of the test equipment after the test, an out-of-calibration condition is noted, the effect of that condition on the test results can then be assessed.

III. DOCUMENTATION AND REVIEW OF RESULTS

Often, test results are processed directly from the test station to a computer data acquisition system and emerge as partially or completely reduced data. Supplemental computer codes may be used to complete the data

RESEARCH AND DEVELOPMENT DIVISION

TECHNICAL PROCEDURE

DATE 12/16/80

1. PURPOSE

The purpose of this document is to define the test procedures that will be used for testing and evaluating cleaners that will be used to remove molybdenum disulfide from the stud-holes of the primary manways on the once-through steam generators (OTSG).

2. REFERENCES

1. Project Technical Plan for OTSG Primary Manway Stud-Hole Cleaning.
2. Technical Procedure ARC-TP-286, Augér Electron Spectroscopy.
3. Procedure for Cleaning of Ferrous Materials, ARC Chem Lab. No. 1201-T52.

3. PRE-REQUISITES

The critical parameters identified in Reference 1 are the weighing of the test assemblies and the application of torque to the test assemblies. The weighing will be done using weights traceable to NBS. The application of torque to the assemblies will be made with a calibrated torque wrench. The final analysis of residual molybdenum disulfide on the stud-hole threads will be made by Augér analysis, performed according to Reference 2. Preliminary cleaning of test assemblies will be performed according to Reference 3.

4. TEST CONFIGURATIONS

One test configuration will be used. This simulates the actual stud-hole configuration and is made by welding a 1/8" carbon-steel plate onto the bottom of a production 2", 8-UN-2 nut. In the tests, a production 2" stud cut in half is threaded into the hole. A second 2" nut is threaded against the bottom nut and simulates the actual man-hole cover. The upper nut is screwed against the stud-hole mockup to a torque of 1450 ±100 foot-pounds to duplicate the tension present on the stud on the OTSG assembly.

5. TEST PROCEDURE

The test will be performed in the following order:

1. Acid clean six sets of stud-holes, studs, and back-up nuts with inhibited hydrochloric acid according to ARC Chem Lab No. 1201-T52 (Reference 3). Rinse thoroughly with de-ionized water and dry with methanol.
2. Weigh each component. (W_1, W_2, W_3)

PROCEDURE NO. ARC-TP-379

PAGE NO. 1

Fig. 9.1. R&D division technical procedure. (Courtesy of Babcock & Wilcox.)

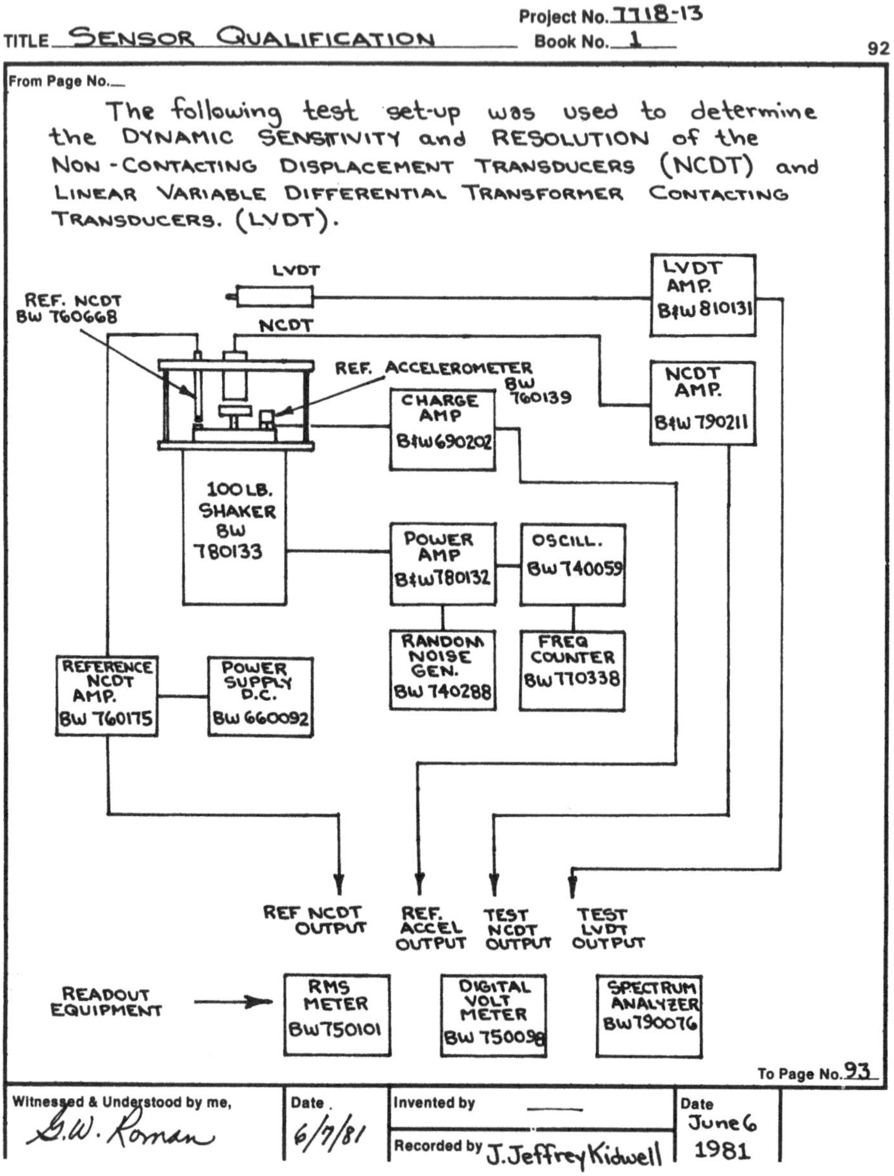
Project No. 7718-13
TITLE SENSOR QUALIFICATION
Book No. 1
92
From Page No.
The following test set-up was used to determine the DYNAMIC SENSITIVITY and RESOLUTION of the NON-CONTACTING DISPLACEMENT TRANSDUCERS (NCDT) and LINEAR VARIABLE DIFFERENTIAL TRANSFORMER CONTACTING TRANSDUCERS. (LVDT).
LVDT
REF. NCDT BW 760668
NCDT
LVDT AMP. B&W 810131
NCDT AMP. B&W 790211
REF. ACCELEROMETER BW 760139
CHARGE AMP B&W 690202
100 LB. SHAKER BW 780133
POWER AMP B&W 780132
OSCILL. BW 740059
RANDOM NOISE GEN. BW 740288
FREQ COUNTER BW 770338
REFERENCE NCDT AMP. BW 760175
POWER SUPPLY D.C. BW 660092
REF NCDT OUTPUT
REF. ACCEL OUTPUT
TEST NCDT OUTPUT
TEST LVDT OUTPUT
READOUT EQUIPMENT
RMS METER BW 750101
DIGITAL VOLT METER BW 750098
SPECTRUM ANALYZER BW 790076
To Page No. 93
Witnessed & Understood by me, G.W. Roman
Date 6/7/81
Invented by
Recorded by J. Jeffrey Kidwell
Date June 6 1981

Fig. 9.2. Block diagram of test apparatus. (Courtesy of Babcock & Wilcox.)

PROJECT TITLE OTSG Primary Manway Stud-Hole Cleaning	REVIEW		
PROJECT NO. 7738-02			
SUBJECT Independent Review	PAGE 1	OF 1	

PURPOSE:

    The purpose of this review is to certify that the acquisition and reduction of the data for the OTSG Primary Manway Stud-Hole Cleaning Test satisfies the project's Technical Plan.

REFERENCES USED FOR REVIEW

- Project Technical Plan for OTSG Primary Manway Stud-Hole Cleaning, Revision 0, 7738-02, November 5, 1980.
- Technical Procedures for OTSG Primary Manway Stud-Hole Cleaning Project, ARC-TP-379.
- Laboratory Notebook, pages 8 through 14.
- Report, LR:80:7738-02:01 - Molybdenum Disulfide Removal from OTSG Primary Manway Stud-Hole - Evaluation of Cleaners.
- Chemical analysis results.
- Auger analysis result sheets.

GENERAL

    The OTSG Primary Manway Stud-Hole Cleaning Program was designed to evaluate the three best cleaning solvents as determined by previous screening tests. The tests were designed to determine the effectiveness of solvents in removing molybdenum disulfide from stud-hole mockups. The cleaning procedures that were used would be adaptable for field use.

    All values (Auger analysis, chemical analysis, and weights) in the report were checked against the references and found to be correct. All calculations in the report were also found to be correct.

    Based on these findings, the results and conclusions reported are accurate. The tests that were conducted satisfy the requirements of the project's Technical Plan.

PERFORMED BY W. S. Leedy _W. S. Leedy_ DATE 12/5/80

REFERENCE SOURCE _____

Fig. 9.3. Independent review of project. (Courtesy of Babcock & Wilcox.)

analysis process. Just as calculations ought to be independently verified
and test instruments periodically calibrated, the computer data acquisition
system should be checked both as an instrument for calibration of the signal
detection amplification and processing and as a computer with appropriate
software verification and validation (see Chapter 10).

The results of experimental testing are evaluated by the project leader
or project engineers under the direction of the project leader. Just as the
calculations for designing the test apparatus should be verified, so should
the calculations for analyzing and interpreting the data being reviewed. The
documentation for the calculations and reviews is similar to that described
in Chapter 5, "Design Control." The emphasis, aside from mathematical
correctness, is on whether the experimental objectives were met and whether the effects and interactions were interpreted correctly. This type of
peer review is practiced in all types of research centers. Discussions on
its use for geoscience projects are provided by Henderson of Sandia Laboratories (4). Oak Ridge National Laboratories used peer review and publication in referred technical journals to satisfy certain QA requirements for
basic research experiments (5). The National Research Council uses peer
review extensively for its publications (6). An example of a review action
is shown in Figure 9.3.

REFERENCES

1. American National Standards Institute, ANSI/ASME NQA-1-1979, Quality Assurance Program Requirements for Nuclear Power Plants, The
   American Society of Mechanical Engineers, New York (1979).
2. American Society for Testing and Materials, Philadelphia.
3. The American Society of Mechanical Engineers, New York.
4. J. F. Henderson, The Application of QA to Geoscience Investigations,
   Transactions, 35th Annual ASQC Quality Congress, American Society
   for Quality Control, Milwaukee (1981).
5. F. H. Neill, Quality Assurance Programs in Research and Development, Quality Assurance in a Large Research and Development Laboratory, Proc. 7th Annual National Energy Division Conference, ASQC,
   Houston (1980).
6. National Academy of Sciences: Entering the Energy Debate, EPR 1,
   Journal, p. 32 (April 1980), Electric Power Research Institute, Palo
   Alto.

Chapter 10

SOFTWARE

I. BACKGROUND

As computers are used in more complex design, computational, monitoring, and control situations, company reputations and company product liability risks become closely associated with the adequacy of computer programs. Particularly in a research laboratory, as the technology advances, so must the means to store, compute, and recall data be improved. Researchers cannot afford the time to perform long manual calculations or to take individual readings at hundreds of measurement points around the test model. This must all be done automatically and efficiently and, therefore, by computer. Yet there is no practical means for proving computer programs correct, particularly those which are developed for one-of-a-kind experiments or tests. There is no universal testing procedure which could be invoked to demonstrate program correctness, e.g., isolated from program requirements and program design structure and documentation. There have been a variety of industry and government investigations to try to solve the problem of how to "test quality in" to computer programs. Huge sums of money have been spent on the development of computer software and, unfortunately, schedule and budget overruns have been a way of life for software development. Often the final products were useless or, at best, something less than was anticipated. In an effort to overcome these problems, rigorous development and documentation procedures were devised. Though costly to implement, the procedures do reduce the element of surprise in software products and improve communications between software developers/users and management.

Computer programs in their final form reside hidden inside the computer waiting for implementation. The information accessed and consequent actions taken by the computer are not easily observed. The possible combination of actions and opportunities for errors often exceed the realm of consideration. Seldom used paths which may be executed in exceptional cases are particularly error prone since the designer may not have envisioned the occurrence of such a situation. Yet, this exceptional case may prove to be the computer's most critical application. (For example, control of critical processes during emergency conditions). These exceptional cases should be planned for and tested if the risk of using computers is to be acknowledged and minimized.

As has been found with hardware, the emphasis in software must be to design and build in quality rather than to rely on testing after the code has been developed. This shift toward defect prevention, rather than detection, is even more critical for software because, once a program is coded, the refinement ceases unless a specific test case is designed for the specific defect contained within the program. The defects designed or built in will not "burn out, " so the hardware product improvement technique of relying on failure over time does not apply for software. A progression of failures into a "bathtub curve" can no longer be used to any advantage to predict how well the system has been debugged.

The latest thrust for resolving this problem, then, is to define the software development process as a set of very distinct phases and institute a process of review through each of these stages or at least those stages for which the highest risk of defect introduction into the program is seen.

II. THE SOFTWARE QUALITY ASSURANCE PROGRAM

Whether software is developed as a support component of a larger R&D effort or as the project end product, the QA considerations need to be integrated into the R&D program structure. The previous descriptions of Project Technical Plans (PTP's) and QA Plans (see Chapter 4) are applicable to software planning, development, and evaluation. In addition to the generic PTP and QA Plan requirements, additional considerations need to be made for software. The questions to be addressed are "What" and "How Much. "

A. Planning

When an R&D project is initiated which requires the use of a computer, the effect of error in the computer results upon the project results should be determined in the initial review. QA standards and procedures should be imposed on software development to assure quality commensurate with project requirements. If the impact of the computer error is great, then a high level of quality assurance must be imposed on software development. If, however, the impact of error is minimal or project results will be reverified later under more rigid conditions, a lower level of quality assurance is permitted.

Prior to any creation of software, the software requirements should be identified and incorporated into a Project Technical Plan, if one is required (see Chapter 4). For standard laboratory practices, generic software development requirements should be documented in a standard procedure. In the absence of any specific Project Technical Plan, this generic procedure would be in effect. Items to be considered in establishing general software requirements for various quality levels are listed below. Note that although the actual amount of quality may be the same because you may have the same type of development requirements for all programs, the difference is in the amount of documentation and overchecking that is performed. The items to be considered apply to all software quality programs, but additional documented evidence must be required based on the level of quality assurance that is required. This should be indicated in the quality assurance plan (see Chapter 4 and Item 13 under Paragraph C).

Software requirements to be added to a Project Technical Plan are:

1. Does the project plan specify the "impact of error" for software to be developed and designate the required level of evaluation and certification?
2. Does the plan specify software cost and schedule constraints?
3. Does the plan specify the required project outputs (include documentation)?
4. Does the plan specify major software development tasks and information flow among activities?
5. Does the plan designate who is responsible for software tasks and outputs?
6. Does the plan identify information required from external sources?
7. Does the plan specify, according to priority, the software characteristics important to the application?
8. Does the plan identify areas of exception to standard implementing procedures?
9. Does the plan specify all required management approvals for project initiation?
10. Does the plan identify external standards, regulatory requirements and customer requirements?

B. Software Quality Levels and QA Plans

Quality levels allow the flexibility to choose a set of development procedures commensurate with project requirements. For example, when preparing a software test plan, several levels of rigor may be used:

1. A formal software test plan is prepared by an independent test team, fully describing the test approach, test environment, test procedures, test cases, expected test results, order of testing and any other information perceived as relevant to the specific testing. The test plan is reviewed and formal approval is gained prior to the beginning of testing.
2. An informal software test plan is prepared by the development team to perform a set of documented tests on the software product. Testing is planned to be a comprehensive process, though innovative, state-of-art testing procedures are not generally required.

3. No test plan is required. Testing may be performed by a test team consisting of the software developer and the R&D project leader. A series of tests are performed and the results are documented. The R&D project leader takes responsibility for the program correctness, since he generally declares testing to be completed.

Large gaps exist between the three levels outlined. Example 3 represents the software testing method most often used today in R&D. These three examples are cited merely to demonstrate that there is wide variation of approaches that can be taken to perform any of the software development activities. Secondly, some variation is appropriate to better balance the quality, cost, and schedule requirements of each project. This variation should be built into the software QA program to permit required flexibility in an R&D environment.

Software requirements to be added to the project QA Plan are:

1. Does the QA Plan identify all applicable implementing procedures (includes procedure date and revision level)?
2. Does the QA Plan correlate software requirements with other QA program requirements to provide guidance to developer?
3. Does the QA Plan specify a software certification and control level?
4. Does the QA Plan specify the organizational independence and technical qualification requirements of the evaluator?
5. Are there provisions for revision level control?
6. Does the QA Plan provide for unique identification of each task output (includes all documentation) by revision level?
7. Does the QA Plan provide a means for approving changes to requirements after they have been initially approved?
8. Does the QA Plan provide a means for establishing current status upon demand during development and during use?
9. Are provisions made for a permanent project record file?
10. Does the QA Plan provide for controlled access to the file during development?
11. Have provisions been made for controlled access to the completed project record file?
12. Does the plan have provisions for storing digital computer programs in a machine readable form?
13. Does the permanent project record file requirements include (ongoing project files should also include):
    Project technical plan
    Project QA plan
    Project Software Configuration Management plan
    Status and progress reports
    All project documentation products
    Reference information from external sources
    Project communication sent
    Project communication received

Project log
Support documentation pertinent to project records or clear reference
to where support documentation can be accessed (e.g., computer
vendor documents)
Directory to record organization
Audit "road map"

C. Software Development Process

The software development process consists of four phases: requirement
specification, design, code, and test. The project technical plan should
have identified organizations responsible for each phase. The persons who
have implemented each phase should also be identified by log entry, progress reports, document authorship, code preamble, or some other means.
The QA plan should identify to what extent reviews are required at the completion of each of these phases. At the completion of the review, discrepancies found should be corrected prior to proceeding to the next phase.
Correction should include revision of all related documents.

The Structured Analysis and Design Technique (SADT) (1, 2), developed by Softech, Incorporated, is a graphic language, capable of adding
structure and order to the development process. Although its simplicity
sometimes masks immediate recognition of its potential power, the fact
that it can be readily understood is one of its greatest strengths. This
technique can be used for a number of applications. It provides the structure and framework for organizing the detail needed for software development; it also permits managers, analysts, design engineers, and others to
relate to software development and also better coordinate their efforts
towards successful project completion. Other techniques exist that provide
the same or similar capabilities.

The constructs of SADT are directed lines and boxes as illustrated in
Figure 10.1.

Both lines and boxes have structural expansion capabilities. An activity defined at a high, abstract level can be decomposed into more detailed
sub-activities on separate pages. The sub-activities can in turn be broken
down into further detail as required for complete, precise definition. The
lines have the same decompositional capabilities. Examples of line expansion into more detail is shown in Figure 10.1. Note in Figure 10.1 that the
USER REQUIREMENTS input to boxes 1 and 2 is further expanded to define
a subset of USER REQUIREMENTS input to box 3, labeled USER ACCEP-
TANCE CRITERIA.

A graphic language presentation similar to SADT is recommended for
consideration when establishing an R&D QA program. QA procedures and
software development methods can be prepared to reference this basic
framework.

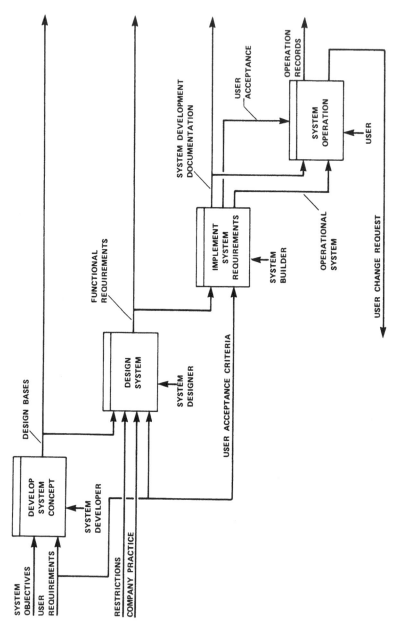

Fig. 10.1. System development process (total picture). [From Reference 3, IEEE. Reprinted, Vol. NS–28, No. 1, p. 915 (February 1981).]

Using the four phase idea discussed at the beginning of this section, checklist items to be considered during each phase are listed below:

Requirements Specification:

1. Does the requirements specification include a description of the problem which the program is intended to solve?
2. Does the specification include a description or reference to the analysis of the required functions to support the adequacy of the chosen problem solution to be implemented?
3. Does it include a complete definition of the functions to be implemented?
4. Does it specify the order in which the functions are to be performed (and relationship between functions)?
5. Does it specify the performance criteria and acceptance criteria for evaluating the program's adequacy?
6. Does it specify the test requirements?
7. Does it specify the development requirements or references where they are specified?
8. Does it prioritize the software quality factors according to relative importance?
9. Does it specify all program inputs?
10. Does the input specification include:
    Data type
    Format
    Units
    Conversion algorithms
    Ranges
    Variable name
11. Does the requirement specification specify all program outputs?
12. Does the output specification include:
    Data type
    Format
    Units
    Conversion algorithms
    Ranges
    Variable name
13. Does the requirement specification identify limitations and exceptions?
    How errors are to be handled
    How errors are to be reported
14. Does it include the required accuracy and timing requirements as part of performance requirements?
15. Does it specify the required or assumed computer environment?
    Identification of the computer and required peripherals
    Operating system requirements or identification
    Required implementation language or language level
16. Is it understandable?

17. Is it consistent in detail and organization?
18. Is it complete? If certain decisions must be postponed until design phase, are they clearly identified?
19. Are the functions structured and identified to facilitate traceability?
20. Does the specification include functional test data to facilitate program functional evaluation?
21. Does it specify user requirements (man/machine)?
22. Does it specify data management requirements?
23. Does it specify any initialization requirements?
24. Does it specify security requirements?
25. Does it specify maintenance requirements?
26. Does it specify availability requirements?
27. Does it specify interface requirements of hardware and software.

   Design:

1. Does the program design include:
   Preliminary design
   Test plan
   Detailed design
   User instructions
2. Is the design documented in a structured, top-down format?
3. Does the preliminary design identify all program components or modules?
4. Does the preliminary design include module interface definitions?
5. Are modules scoped in the preliminary design to perform only one primary function?
6. Is the preliminary design traceable back to the requirements specification?
7. Does the preliminary design specify the principle data structures?
8. Is the detailed design traceable to the preliminary design?
9. Is the detailed design sufficiently detailed to support coding without other references?
10. Does the detailed design include provisions to support module and total program testing?
11. Does the detailed design identify and define all data?
12. Are the detailed data definitions traceable back to the preliminary design data structures?
13. Does the design include a complete definition of the program's external interfaces?
14. Is the design complete? Are all required functions included?
15. Does the test plan include:
    A test approach
    Test environment description
    Test procedures
    Test cases with expected results
16. Does the test approach specify a test strategy for the required test coverage?

17. Does the test environment completely define the test set-up and supply sufficient information for an independent team to perform testing?
18. Are test procedures sufficiently defined for an independent person to perform all defined tests?
19. Does the test approach and supporting test cases support functional testing and program design testing?
20. Are test cases included for the unusual unplanned inputs in addition to typical input sets?
21. Does the test strategy include provisions for bench marking against experimental results or other programs? If so, are the bench marking data documented for future reference?
22. Can the design be analyzed for preliminary evaluation against performance criteria?
23. Does the test procedure reference user instructions?
24. Can user instructions be followed by an independent user not previously knowledgeable of the program?
25. Do user instructions give directions for recovery when an error occurs during operation?
26. Do user instructions include installations and initialization instructions?
27. Are periodic tests furnished for diagnostic purposes?

Code and Debug:

1. Is the code consistent with the design? Do the same module and interface definitions exist?
2. Can the code be traced back to the design and the requirements specification?
3. Does the code include a preamble for each module which identifies:
Name of code developed
Revision history by number and date
Purpose and function performed by the module
Description of all inputs, outputs and internal variables
Data type specification of variables
Comments to support understandability and traceability to design
4. Does the code have provisions for sequence identification (smallest unit to be controlled must have a unique identifier)?
5. Are errors uncovered during debugging recorded?
6. Is a clear distinction made between the end of debugging and beginning formal testing?
7. Is the code, once debugging is completed, uniquely identified by revision level for positive identification for test purposes?
8. Is code structured in accordance with the design?
9. Are structured coding practices followed?

Test:

1. Is testing performed in accordance with the test plan?
2. Are test results documented for analysis and future reference?
3. Are test results traceable to test cases specified in the test plan?

4. Are discrepancies identified during testing corrected in code, design, and even requirements specification if affected?
5. Is a test log kept to record observations during testing?

D. Evaluation

Reviews should be conducted at the completion of each major activity. The frequency and thoroughness of formal reviews is dependent on the level of quality assurance required. In addition to required reviews, a means must be provided to evaluate the output and, perhaps, the performance of each completed activity. The appropriate review criteria must provide a way to answer the questions: How completely are the software functions defined? Do the software requirements adequately address the required functional performance? Does the installed program faithfully represent the specified functions? The questions to be answered should be established during the planning stage so that activity performance and documentation can be best organized to allow the reviews to be more easily accomplished.

All items which were considered and selected for inclusion during the planning phase from the checklist under Sections II. A, II. B, and II. C are included as part of the evaluation checklist. Additional checklist items may be drawn from the following list:

1. Do software project records include (as a minimum):
   Requirements specification
   Design
   Code
   Test results
   Programmer's log
   User instructions
2. Does program documentation meet the following (recommended minimum for any software development):
   Is it legible and understandable?
   Is it organized so that an independent person can read and understand it?
   Will it support future project documentation?
   Are any independent reviews held properly documented?
   Does the programmer's log provide a chronological record of the programmer's experience, decisions made, pertinent project events?
   Are project records secured to minimize loss (loose-leaf notebook, comb bound or other)?
   Is a copy of all records secured in permanent records file and a copy available to support future maintenance?
3. Is the designated level of independence met for all required evaluations?
4. Does a Certification Approval form exist for the computer program?

5. Does the Certification Approval form include:
   Date
   Program identification number, including revision level
   Program title
   List of documentation applicable to certification review
   Specified "impact for use" category
   Abstract
   Appropriate quality assurance statement
   Cognizant engineer's signature
   Developer's signature
   Independent reviewer signature
   Certification coordinator
6. Have all programs to be used met specified "impact of error" require-ments for this application?
7. Does all required program documentation exist?
8. Does a plan exist to upgrade the program if necessary?
9. If the program cannot meet specified requirements, has the necessary exception approval been received?
10. Do project records include all documentation and certification state-ments or reference to where they can be accessed?

QA can provide a mechanism for creating management awareness and for establishing adequate procedures to minimize risks associated with computer applications. QA provides a link between management and soft-ware technical specialists to establish and maintain a desirable balance be-tween cost, schedule and quality to best serve company goals and objectives. Obtaining management's commitment to invoking procedures to control soft-ware development review and documentation is difficult because we do not know how to demonstrate to management, quantitatively, how a cost benefit relationship of X amount of verification and review at the front end saves a Y amount of dollars for revision socts or code failures at the back end. We are forced to rely on management's intuitive insight to recognize that prob-lems will exist or wait until some catastrophic accident occurs to point out the problem. While the latter course may be the most effective in driving the point home, it is certainly not the most responsible one.

REFERENCES

1. Douglas T. Ross, Structural Analysis (SA): A Language for Communi-cating Ideas, IEEE Transactions on Software Engineering, Vol. p. 00 (January 1977).
2. Softech, Inc., An Introduction to SADT, Softech Document #9022-78 (February 1976).
3. N. C. Thomas, Incorporating Software Into Nuclear Power Plant Sys-tem Design, IEEE Transactions on Nuclear Science, Vol. NS-28, No. 1 (February 1981).

Chapter 11

RECORDS AND REPORTING

I. PROJECT RECORDS

The minimum records considered essential for documenting research work are discussed in Chapter 3. Records should include any information, memos, meeting minutes, instrumentation calibration curves, or special references that have a bearing on the direction the project has taken, the data acquired, or the interpretation of results. These would normally include the following: work orders and customer communications, purchasing documents, test article configuration, laboratory notebooks, and reports. To this can be added the specific records committed to by the project leader to meet quality assurance requirements and that are listed in an appropriate section of the quality assurance plan (see Figure 11.1). As a general guideline, if a commitment was made by the research center (project leader) or to the research center, that commitment should be documented in the project files and compliance with that commitment should be also documented.

II. REPORTS

The reports published by a research and development (R&D) facility are the principal evidence of the quality of its performance and ability. Since personnel at the R&D center are judged by the contents of these reports, their writing ability must match their research ability in the laboratory. The production of effective reports, therefore, is a major part of professional performance.

"Letter" reports are intended to promptly transmit detailed results of a project to technical counterparts in operating divisions, and to permanently record data, operating procedures, equipment descriptions, etc., in sufficient detail to be understood by present and future technical specialists on the subject. The format of a letter report is not rigid. Illustrations and tables are appropriate. The illustrations can be prepared by a graphics group or sketched carefully by the author, but should be neat enough for

QA PLAN — PAGE 3 OF 3   PROJECT _____   _____

SECTION	REMARKS
17.0 QA RECORDS	
RC-194 ☐	
GO 4028 - 4029 ☐	
IWO ☐	
PROPOSAL ☐	
PROJECT TECHNICAL PLAN ☐	
QA PLAN ☐	
INDEPENDENT TECHNICAL REVIEW ☐	
ROUTE SHEETS ☐	
TECHNICAL PROCEDURES ☐	
INSPECTION CHECKLISTS ☐	
AUDIT REPORTS ☐	
CORRECTIVE ACTION REPORTS ☐	
MATERIAL I. D. TAGS ☐	
DISCREPANCY TAGS ☐	
LIST OF INSTRUMENTS ☐	
TEST LOG ☐	
TEST DATA SHEETS ☐	
CALCULATIONS/REVIEWS ☐	
COMPUTER DATA/REVIEWS ☐	
LIST OF DRAWINGS ☐	
SUPPLIER QUALITY HISTORY ☐	
SOURCE INSP. REPORTS ☐	
PURCHASE ORDERS ☐	
INTERIM REPORTS ☐	
DESIGN REVIEW REPORT ☐	
FINAL REPORT ☐	
OUT OF CALIBRATION REPORTS ☐	
10 CFR 21 REPORTS ☐	
18.0 QA AUDITS	
INTERNAL ☐	
SUPPLIER ☐	
INFORMAL (SURVEILLANCE) ☐	
19.0 CONTROL OF CUSTOMER	
FURNISHED PROPERTY ☐	

APPENDIX A
DOCUMENTS REQUIRING CUSTOMER APPROVAL
DOCUMENT   APPROX. SUBMITTAL SCHEDULE

Fig. 11.1. QA plan index of project records. (Courtesy of Babcock & Wilcox.)

clear understanding. Conclusions and recommendations can be omitted or partially covered, depending on progress or on requirements in the case of contractual or customer sponsored work.

There is no limit to length. However, a lengthy report probably indicates the need to subdivide, and a summary is necessary at the beginning of the report. To best communicate to most readers, a summary is needed when the report contains more than three pages of narrative.

"Formal" reports are intended to communicate the significant results, conclusions and recommendations of an R&D project, or major part thereof, and are written for a broad spectrum of readers, particularly middle and upper management. They should be prepared annually, or at the completion of a major work phase, whichever occurs sooner. They are comprehensive summaries of the projects.

An introduction, results, discussion, conclusions, and recommendations are applicable to every formal report. Additional sections and/or arrangement are optional. Exhibits should be designed to communicate general ideas, and graphs and curves need not be readable with high precision. Appendices (perhaps in a separate volume of limited distribution) are preferred for voluminous tables of data and results. All pertinent "letter" reports prepared during the project should be included as part of the appendices.

Most beginning writers tend to order their material on a chronological or problem solving sequence, just as they organize the research project itself. Unfortunately, this arrangement emphasizes the methods and equipment used, rather than the implications of the research. It is preferable, therefore, to alter the sequence with emphasis on giving management what they want most to know. In this case, the first sections would be introduction, results, discussion, conclusions and recommendations, followed by the remaining supporting sections and appendices. This outline requires more writing skill, since the details of apparatus, procedure and analysis have not yet been introduced. However, once learned, the continuity of thought from problem through objectives, results, conclusions and recommendations is preserved without interruption — closely knit thoughts that can be beneficial to both writer and reader.

III. TRACEABILITY

It is very helpful if a unique job or project number is assigned and used to mark each of the records. This allows them to be quickly collected and sorted for future reference. For large, long-term test programs involving many independent test phases, separate QA plans may be necessary to adequately identify the needed controls. If this is the case, it is recommended that the job traceability number be likewise modified so each QA plan has a unique number. Since the QA plan specifies the project records, the project phases can be accurately documented. Numbers are often assigned to formal research reports, but the reports are usually written at the completion of the project. Marking of records should begin when the records are generated.

IV. RETENTION

The retention of records is of concern because there is always the question
of, "how much should be kept and for how long"? Nuclear requirements
specify retention for the life of the plant, which is usually assumed to be 40
years, but could go on much longer depending on the time and effort required
for complete decommissioning of the facility. ANSI/ASME NQA-1 (1) describes typical lifetime records, among which are 13 types directly involved
with the design. Since the results of R&D are most often used to affect the
design of hardware or systems, these document retention requirements
should be looked at carefully. The application of R&D to the manufacturing,
installation, inspection and operation is not intended to be down-played, however, and research reports having a significant bearing on these areas must
also be considered for long term retention.

     Not all R&D activities have the same regulatory environment which
spawned the standards for controlling every aspect of nuclear design, construction and use. But the potential for damage to the public and resultant
litigation against the designer, manufacturer, or the research organization
providing a service to them is much greater in the non-nuclear field because
of the general complacency toward these seemingly safer items offered in
the marketplace. Automobiles have been shown to be more dangerous than
nuclear power plants, and expensive recalls are almost commonplace, but
there are few, if any, standards or guidelines for R&D. The background
and thought put into nuclear standards is extensive and much of the rationale
is applicable to any product line.

     Generally speaking, research records should be retained as long as
any of the vital records for a research center. If the data pertains to a
specific product, process or facility, the records should be retained as
long as that entity is in use. Results of basic research should be retained
as long as the research organizational entity exists.

REFERENCES

1. American National Standards Institute ANSI/ASME NQA-1-1979, Quality
    Assurance Program Requirements for Nuclear Power Plants, American
    Society of Mechanical Engineers, New York (1979).

Chapter 12

AUDITING THE PROGRAM

I. GENERAL AUDITING CONSIDERATIONS

Effectiveness of R&D is greatly influenced by the effectiveness of control systems used within the R&D organization. Control has been singled out as one of the more important but difficult problems of R&D management (1). The Hughes Aircraft study report on R&D productivity points out:

> ...the existence of a control system does not guarantee that control exists, nor does adding more controls necessarily result in better control. It is the quality not the quantity of control systems that is important. Therefore, control systems must be frequently and critically appraised to ensure that they remain effective.

Wachniak (2) describes a management-oriented, participative approach used to audit research and development activities within operating divisions of a company. The theory of the audit is to review these activities from the perspective of the management functions of Planning, Organizing, Directing, and Controlling, and a series of internally generated standards. The audit approach is centered on auditee participation with the objective of developing a problem-solving partnership instead of the more traditional active auditor-passive auditee relationship. The auditee is expected to respond to a questionnaire and supply documentation indicating whether the standards agreed upon for research and development activities are being met. Where necessary, an assurance audit is conducted.

Systematic internal audits assess the degree of compliance with internal direction given the employees of the research and development division. This evaluation should determine whether personnel are currently informed of the procedures that affect their activity. These may include safety, government codes and product liability considerations, as well as requirements for documentation and verification.

Internal auditing may require an intensive evaluation of all aspects of one specific project using the QA plan and project technical plan as a basis for tailoring the checklist. Another method is by statistical evaluation of a

large population using a standard checklist which incorporates the requirements of a few relatively stable procedures. The statistical method is best
for systems (e.g., purchase order processing or measurement control) or
large numbers of projects with no specific QA or project technical plan requirements.

Regardless of the method used, there is a common thread of audit
technique to be applied. Audits should be conducted on a periodic basis,
usually annually, or at least once during the course of a project for projects
with specified QA. Additional audits may be scheduled if warranted by evidence of a quality problem. The auditing process begins by establishing a
schedule of the organization or function or project to be audited.

Audit team leaders should be selected to direct and coordinate the activities. The team leader then makes an initial contact with the involved
section manager or project or activity to be audited. The team leader then
selects the participating team members. Audit team members should be
familiar with the process or project being audited, but should not have any
direct responsibility for the activity to be audited. Members may be re-
cruited from organizations other than the QA organization to provide expertise in specific categories to be audited.

The audit team leader then reviews the reports of previous audits of
the activity conducted by the R&D QA organization or by customers and selects the areas to be audited from the applicable procedures. It is common
courtesy to provide at least one week advance notice to the appropriate personnel, giving the planned audit dates, the names and organizations of the
audit team members, and the general listing of the audit categories to be
evaluated. At that time, it would be appropriate to request personnel from
the activity being audited to accompany the audit team members during the
audit. These personnel can be used to clarify and amplify areas under
evaluation when necessary.

A meeting of the audit team members should be scheduled during the
week preceding the actual audit. The team leader will assign the audit
categories to the team members and provide a copy of the appropriate procedures and audit checklists. During the meeting, the team leader should
inform the auditors of previous system violations and the corrective action
taken by the activity to be audited. It is then up to the team members to become familiar with the contents of any regulatory codes or specifications,
plans and procedures which govern the project or the system to be audited,
and also to become familiar with any previously reported discrepancies and
the corrective action which had been committed.

During the first day of the audit, the audit team should meet with the
section manager or group supervisor or project leader of the activity to
describe the purpose and scope of the audit, and begin the audit. The team
members should proceed to their assigned work areas and ask the appropriate questions from the audit checklist and to record the answers to the questions. The checklist should be used as a general guide, but if additional

areas of investigation are indicated based on the areas of the checklist questions, a team member should not hesitate to pursue those additional areas, regardless of whether they are actually listed on the audit checklist. Answers to the audit questions should all be verified by reviewing appropriate procedures, recorded data, qualification records and approvals or test reports, or by actually witnessing the operations or inspections. The team members should also record any verification and references, such as procedure numbers, drawing numbers, serial numbers, route sheet numbers, or test reports. The team member evaluates the effectiveness of the corrective actions taken for previously reported findings. Any recurrence of a previously reported finding should be again reported, with a specific note that it is a repetitive item.

The team leader reviews the audit findings and recommendations with the team members prior to the post audit critique meeting to be sure there is an agreement with the findings and recommendations. Then, he arranges to have the audit findings typed and reproduced, if possible, prior to the critique meeting.

On the last day of the audit during the post-audit critique meeting, each team member presents findings and recommendations orally, and should be prepared to answer any questions or provide any clarifications. At this time, it is then desirable to have the attendees all agree that they understand what the audit finding is and that the facts presented were true facts. It may not be possible for everyone to agree that it constitutes an actual audit finding or a violation of a particular specification, but there should be a general agreement that the facts are being presented in an objective and truthful manner.

It is usually the responsibility of the team leader to prepare the final report, and again, it is customary that within a reasonable time, for example, two weeks after the critique meeting, one or more copies of the audit report should be transmitted to the section manager, group supervisor, or project leader of the audited activity. The transmittal letter accompanying the audit report should outline the criteria for responding to the audit findings. It is customary to allow a reasonable length of time, for example, ten working days, for the audited personnel to issue a response to the audit findings. The group being audited should investigate and respond to the audit findings as promptly as possible, giving specific evidence of corrective action or of a firm commitment, time schedule for implementation of corrective action.

The audit replies are forwarded back to the audit team leader, who then distributes them to the team members who initially reported the findings. The team members then evaluate the replies and proposed corrective actions and forward their comments to the team leader. Any replies deemed to be unacceptable should be handled by the team leader. The team leader assures all findings have been satisfactorily resolved and indicates this in a memo to the audited activity. The transmittal of the final audit closeout letter should include distribution for the original audit report.

Final verification of implementation of the committed audit actions may be held off until the next regularly scheduled audit if the findings are relatively minor, or they may be the subject of a special follow-up action or reaudit if they are of a major nature.

II. PROJECTS

A. Specified QA

Having proceeded through the general scenario for performing an audit, we should now look at some of the differences which arise when performing the various types of internal audits. As stated before, one approach for performing an internal audit is the intensive evaluation of all aspects of a specific project using the QA plan and project technical plan as a basis for tailoring the checklist. The QA plan identifies the specific implementing procedures which are expected to be encountered during the course of the project. Standard audit checklists can be developed for those procedures and used as they are encountered throughout the project, but the audit checklist must take into account more than just the standard procedures. It must recognize that the project technical plan has called for certain tasks to be done which the administrative procedures obviously could not foresee. Therefore, in the case of test control, for example, the audit checklist would require examining the test procedures to make sure that the appropriate forms were used, signatures were applied, and the distribution and document control requirements were adhered to. The administrative procedures could not take into account, however, what types of tests were to be run, at which pressures or temperatures or how many runs were to be taken at each level. That is specified in the project technical plan. The audit checklist must include questions pertaining to verifying that these tests were run at the various committed parameters. The completed audit then verifies that the project was conducted, not only in accordance with the committed procedures, but that the technical aspects in the project technical plan have also been complied with.

Audits of projects with specific QA requirements are concerned with the assurance that those requirements have been complied with. Therefore, the findings resulting from those audits would be expected to be corrected soon after the conclusion of the audit. For that reason, the appropriate corrective action reporting forms would be incorporated as a part of the audit report, and specific commitments for corrective action required from the project leader.

The audit report for project audits with specified QA can take advantage of the necessity for corrective action report forms, quality plans, and formal checklists. The report can be streamlined because the quality plans and project technical plans act as work sheets to back-up the audit report. Often if project audits are repeated, it is not necessary to bring the reader of the audit report up to speed as to why the audit is being conducted and what areas were reviewed. This was already identified by the QA plan, the project technical plan and in a formal written checklist which is included in the project audit files.

To aid the reader in understanding the highlights of the report, a summary is given at the beginning of the audit report, which states when the audit was conducted, a general qualitative evaluation of the project and a listing of significant discrepancies. If the reader desires, he may then read the rest of the report, which gives the specific requirements, project technical plan and QA plan numbers, and can go into more detail as to the findings and recommendations from the audit. Finally, if there are any specific corrective actions that must be taken after the conclusion of the audit, they can be documented on appropriate corrective action reporting forms and attached to the audit report. As can be seen, this type of reporting is best suited where sufficient back-up detail is kept in the files and not presented in the body of the report.

B. Standard Laboratory Practice

The auditing and reporting techniques for projects performed under "standard laboratory practice" are geared not so much to verifying compliance with a set of pre-planned QA requirements, but to obtain an overall characterization of how the projects are being performed with respect to general research center guidelines. Therefore, the checklist for the project audits is taken from the requirements of the various administrative systems encountered during the course of the project. In this type audit, a statistical sample of projects is taken from the total number of projects conducted during a specific time (e.g., fiscal or calendar year). A checklist of standard questions that could be asked of any project is then generated and, as each project is reviewed, if a specific supporting system, such as the procurement document control system or measurement control system, is used, the checklist developed for each of those systems could be employed. When making inferences as to what degree the individual systems are being complied with, one has to take into account that although the sample of projects audited may be high enough to give, for example, a 90% confidence with respect to the total population of projects, only a small fraction of that sample may have involved purchase orders. Yet there is a large population of purchase orders generated at the research center. Obviously, the smaller sample size gives a smaller degree of confidence as to the inferred defect rate within the purchase order system.

The reporting of audit results for these type of projects is somewhat different from those projects for which there are specified QA requirements. Because of the sampling nature, and because there are no specific QA requirements to be certified, a recommended approach would be to take a large sample and plot the number of defects found within that sample and then let the data speak for itself, rather than trying to identify each defect on an individual corrective action report and then attempt to obtain specific corrective action within each project for each defect. By summarizing the defects statistically, the reader can make up his own mind as to how important it is for the item to be corrected and to what extent the defect is likely to exist within the overall population.

A potentially serious limitation to attempting to collect audit data statistically is that the checklists are written in a "yes" or "no" format. If the auditor follows that checklist blindly, good statistics might be obtained, but the opportunity to uncover and resolve serious problems might be lost. Any checklist is a guide. The auditor should feel free to leave the checklist and pursue an independent line of questioning when answers given appear to warrant that action. The human senses can pick up nonverbal reactions or respond intuitively to dynamic situations in ways never anticipated by the checklist writer. If the leads are not fruitful, the auditor should then return to the checklist.

Because there may be some justification required as to sample size and auditing techniques, a different format of audit report should probably be employed than the boiler plate format described in the preceding section. In this case, the format would be more closely resembling that of a "formal" report (see Chapter 11). The contents of this type of audit report would include typically an introduction, results, discussion, conclusions, and recommendations.

III. SYSTEM AUDITS

The generic system audit can be handled much in the same way as the project audits for standard laboratory practice, in that for major systems there are generally sufficient numbers of examples of documentation to be examined that it would be impractical to do a 100% evaluation of all the documents. Therefore, statistical sampling would be employed to obtain some inference as to the degree of compliance with the requirements for that system. In this case, however, the system would probably involve a fewer number of procedures than is likely to come up with the project audits, and because of the highly repetitive nature of the audit, and consistent application of the procedural requirements, a checklist can be more readily prepared which would cover all aspects of the procedures and lend themselves to a better statistical inference of the degree of compliance. The audit approach and reporting would be same as previously discussed under the general section of this chapter, and the report format would be the same as discussed under Section V.B. Bar charts are valuable tools for displaying degrees of compliance within the procedures of a given system, because a bar chart like the one shown in Figure 12.1 can display the procedure bro-

REAUDIT OF CATEGORY 1 — PERCENT DEFECTIVE BY AREA CHECKED

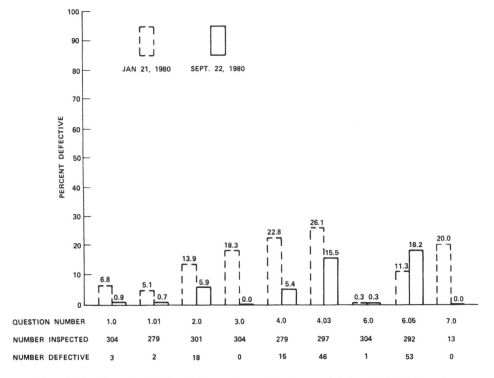

Fig. 12.1. Bar chart of audit results. (Courtesy of Babcock & Wilcox.)

ken down into specific checklist questions (see Figure 12.2). The times a
question receives a "no" answer can be shown as a percentage of defects
against that given question. Statistical analysis is kept to a minimum, but
the message is clear as to where the problems are. As with audits of
standard laboratory practice projects, it is more effective to take a large
sample, show a significant trend of defects, and allow the numbers to sug-
gest the appropriate course of action, rather than seizing upon what may be
a random defect and pressing for immediate corrective action of that one
area. Naturally, the variations and approach toward corrective action are
dependent upon the seriousness of the defects found. It would definitely not
be in a company's best interest to allow any really serious defect to go un-
corrected just for the lack of having sufficient statistical data to prove that
a trend exists. This is where the auditor's judgment and experience take
over from the unbiased objectivity of pure numbers.

Checklist for Category 1 Instruments

Follow-Up Audit of 9/22/80 Only

1.0 Locate part and verify correctness of following data (compare part with computer run).

    1.01 Company Serial Number

2.0 Verify part location is per computer run.

3.0 Verify part is assigned to Section identified by computer run.

4.0 Verify part has a "certification" label applied.

    4.03 Is calibration current at time of audit?

6.0 Verify service log file has been established for correct serial number.

    6.05 Model Number

7.0 For out of calibration conditions, verify out of calibration report has been initiated per Administrative Procedure 1713-02.

Revision Number and Date: 00   September 16, 1980
                        (Number)   (Date)

Approved:

Fig. 12.2. Sample audit checklist. (Courtesy of Babcock & Wilcox.)

IV. SUPPLIER AUDITS

Suppliers can be a significant factor in the quality program for a research center, particularly suppliers of instrumentation or calibration services (see Chapters 6 and 8). Techniques for auditing these suppliers are similar to those for auditing practically any other type of supplier for R&D or production procurements. Helpful suggestions and instruction in performing supplier audits is given by Johnson (3). Another extremely helpful guide is the Auditor Training Handbook published by ASQC (4). Since the mechanics of performing supplier audits are not particularly unique for the R&D application, the reader is advised to consult any of these texts.

REFERENCES

1. Hughes Aircraft Company, R&D Productivity, Second Edition, Hughes Aircraft Company, Culver City, CA (1978).
2. R. Wachniak, Participative Audit — A New Management Tool, International Conference on Quality Control, Tokyo (1978).
3. L. M. Johnson, Quality Assurance Program Evaluation, Stockton Trade Press, Inc., Santa Fe Springs (1974).
4. Nuclear Quality Systems Auditor Training Handbook, American Society for Quality Control, Milwaukee (1980).